AA003154

Energy and Transport Processes 2013

Core Programming Area at the 2013 AIChE Annual Meeting: Global Challenges for Engineering a Sustainable Future

San Francisco, California, USA
3 – 8 November 2013

ISBN: 978-1-63439-038-5

Printed from e-media with permission by:

Curran Associates, Inc.
57 Morehouse Lane
Red Hook, NY 12571

Some format issues inherent in the e-media version may also appear in this print version.

Copyright© (2013) by AIChE
All rights reserved.

Printed by Curran Associates, Inc. (2014)

For permission requests, please contact AIChE
at the address below.

AIChE
3 Park Avenue
New York, NY 10016-5991

Phone: (203) 702-7660
Fax: (203) 775-5177

www.aiche.org

Additional copies of this publication are available from:

Curran Associates, Inc.
57 Morehouse Lane
Red Hook, NY 12571 USA
Phone: 845-758-0400
Fax: 845-758-2634
Email: curran@proceedings.com
Web: www.proceedings.com

TABLE OF CONTENTS

(35a) A Study of Channel Dimension Effect On Performance of Interdigitated Flow Field PEM Fuel Cells

Santamaria, A.,

Cooper, N. J., Brown University

Becton, M.,

Park, J. W.,

Originally presented on: 11/4/2013 8:30:00 - 8:52:00

Novel water management and reactant distribution strategies are important to the development of next generation polymer electrolyte membrane fuel cell systems (PEMFCs). Over-hydration (flooding) or under-hydration (membrane drying) of channels and critical membrane electrode assembly components can have detrimental effects on cell performance and lifetimes. Improving these strategies in PEMFCs leads to higher power density and reduced stack size for vehicle applications, which in turn reduces weight and improves the price competitiveness of fuel cell systems.

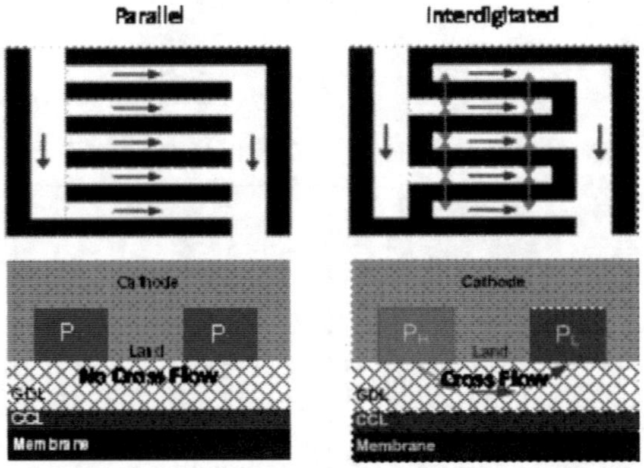

Parallel and interdigitated flow fields, Figure 1, are two common types of PEMFC designs that have benefits and drawbacks depending upon operating conditions and flow-field geometry. Parallel flow fields suffer from longer diffusion lengths which inhibit delivery of reactants and removal of byproduct water. Interdigitated flow fields induce convective transport, known as cross flow, through the porous GDL between adjacent channels and therefore are superior at water removal beneath land areas, which can lead to higher cell performance. Unfortunately, forcing flow through the GDL results in higher pumping losses as the inlet pressure for interdigitated flow fields can be up to an order of magnitude greater than that for a parallel flow field. Additionally, the pressure gradient between inlet and outlet channels in an interdigitated flow field may not be evenly distributed along the channel length. This may cause cross flow maldistribution in the same way Z-manifolds may unevenly distribute gasses across the system. Such maldistribution of reactant gasses along the active area of a cell may lead to areas of low cross flow, and areas of excessive cross flow. This, in turn, can cause areas of low oxygen concentration, water build up, which can lead to higher pressure losses and uneven membrane hydration all of which reduce overall cell performance. This research seeks to examine the affect varying channel dimension sizes have on the performance of a fuel cell under interdigitated

Figure 1: Interdigitated and parallel diagrams. Inlet channels: P_H (high pressure), Outlet channels: P_L (low pressure)

Figure 1: Interdigitated and parallel diagrams. Inlet channels: P_H (high pressure), Outlet channels: P_L (low pressure)

conditions. The features of interest are channel length, width, and depth. The width of lands for the bipolar plates mirrors the width of channels on that plate. Changing feature size affects the distribution of liquid water, as well as the distribution of gaseous reactant in the cell. This work will help develop a better understanding of cross flow for interdigitated flow fields.

For the length study portion of this body of work, a PEMFC was designed that could be configured to run under varying interdigitated channel length configurations through the use of separate manifold exit ports. The lengths studied were 5 cm, 15 cm and 25 cm each with a width of 2 cm. The flow field is comprised of 10 channels with width/depth/land dimensions all of 1 mm. The cell was controlled using stoichiometry to account for the change in active area at the different channel lengths. Tests were performed at stoichiometric conditions of 1.5 on the anode and 2.0 on the cathode (normal), and 2.0 on the anode and 4.0 on the cathode (high). Humidity, temperature and cell compression were held constant for all trials. Pressure was measured at the inlet of the cell while the outlets were exhausted to atmosphere. The cathode

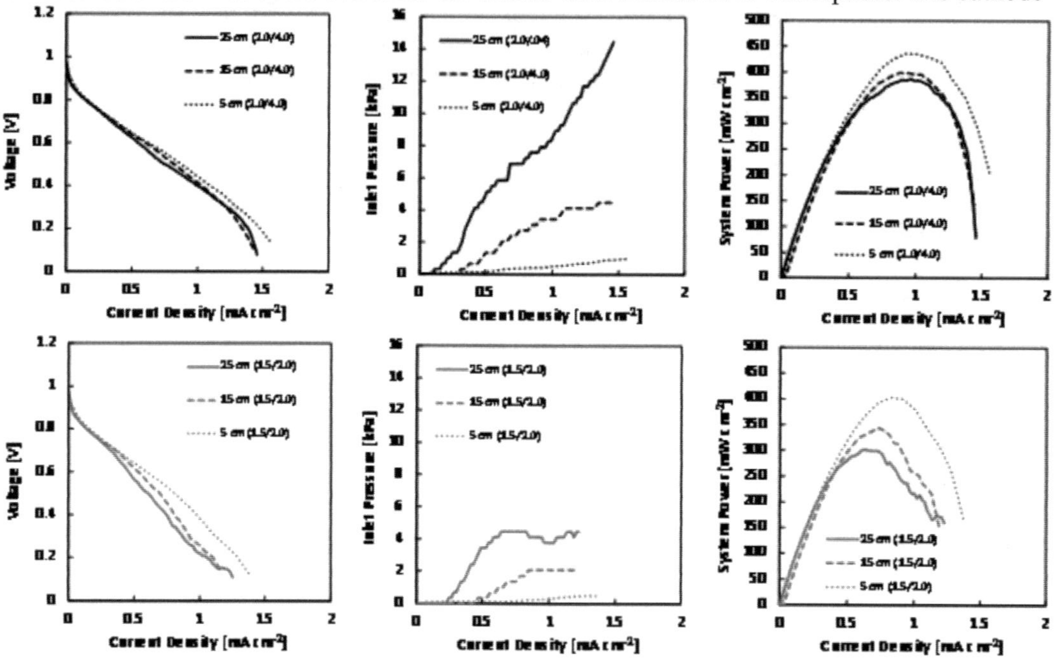

was run in interdigitated configuration, while the anode was run in parallel to isolate the effects of the cathode. Polarization and inlet pressure curves were produced for each length case at the two different stoichiometries. A total of three trials for each case were conducted in random order so as to verify repeatability. Figure 2 displays the average polarization, inlet pressure and net system power data, taking into account pumping losses, for each different length. Reduced cell performance with increasing channel length is a trend that was observed.

This experimental work was complemented by a modeling study of the reactant flow through the GDL at these various lengths and flow rates using the multiphysics package COMSOL. Using the dimensions from the experimental cell, interdigitated flow fields at varying lengths and conditions were simulated in three dimensions. The model focuses on the cathode flow field and GDL layer and does not take into account the electrochemical reaction. As this study is interested in how channel length affects pressure and cross flow distribution, modeling the entire electrochemical process is not necessary since the electrochemical reaction has little effect on bulk cross flow; additionally, the experimental portion of this work characterizes cell performance. Reaction by-product water is a factor;

Figure 2: Polarization, inlet pressure (cathode) and net system power for each length test case. The top graphs are for the high stoichiometry condition (2.0/4.0) and the lower are for the normal condition (1.5/2.0).

Figure 2: Polarization, inlet pressure (cathode) and net system power for each length test case. The top graphs are for the high stoichiometry condition (2.0/4.0) and the lower are for the normal condition (1.5/2.0).

however, no models can currently account for this completely. Figure 3 displays some of the key results from the modeling work. The pressure distribution in inlet and outlet channels was shown to be more variable in the 25 cm cell than in the 5 cm cell. As a result, the cross flow between adjacent channels in the long cell is maldistributed along its length compared with the short cell, which shows a very consistent cross flow rate. This effect may help explain why the shorter channel length configuration produced the best performance. The center region of the 25 cm cell, where the cross flow rate is lower, may be subject to reduced oxygen concentration and increased water buildup. Regions toward the entrance and exit, while receiving higher cross flow rates, may be oversupplied and contributing to increased pressure drop due to higher velocities.

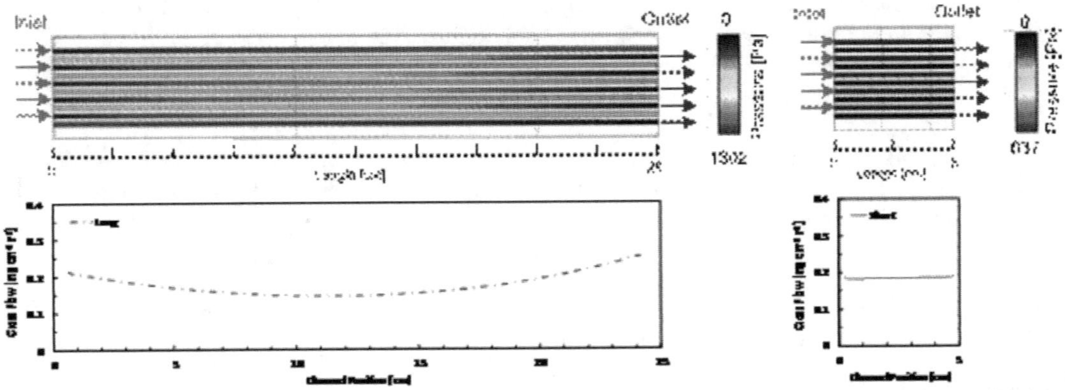

The even cross flow rate in the short cell appears to be more advantageous when these results are compared to the experimental data and suggests that interdigitated designs are sensitive to maldistribution along the channel length. The results of this work were compared to a similar earlier study on parallel flow field channel length; the overall trends are summarized in Figure 4. These findings suggest that for large aspect ratio fuel cells (e.g. long channel lengths and small channel widths) if interdigitated designs are attempted, attention must be paid to the distribution of reactant gas cross flow. Compared to parallel flow fields, which follow a reverse trend due to increased slug removal with longer channel length, interdigitated design flow patterns should be oriented so channel length is minimized. Neutron radiography is being conducted to study the water removal characteristics of interdigitated

Figure 3: (Top) Pressure distribution in long and short interdigitated flow fields. (Bottom) Distribution of cross flow rate along the length of the cell.

Figure 3: (Top) Pressure distribution in long and short interdigitated flow fields. (Bottom) Distribution of cross flow rate along the length of the cell.

cells of varying length. Through-plane images of in-situ cell operation should allow for comparison of inlet and outlet water content as well as comparison of GDL water distribution.

The affect of width and depth of channels are currently being examined. Channel dimensions of interest are (width x depth, in millimeters) 1x1, 0.5x0.5, 0.25x0.25, 0.5x1, and 0.25x1. The effect of these features is being explored through the use of interchangeable bipolar plates. The bipolar plates are 20 cm long and 1.5 cm wide, for a total active area of 30 cm^2. The bipolar plates are machined out of aluminum and have a nonreactive gold coating on the active areas to prevent corrosion and eliminate unaccounted for reactions on the surface of the plate. The use of gold further reduces the interfacial contact resistance with the GDL, decreasing the voltage loss due to ohmic resistance.

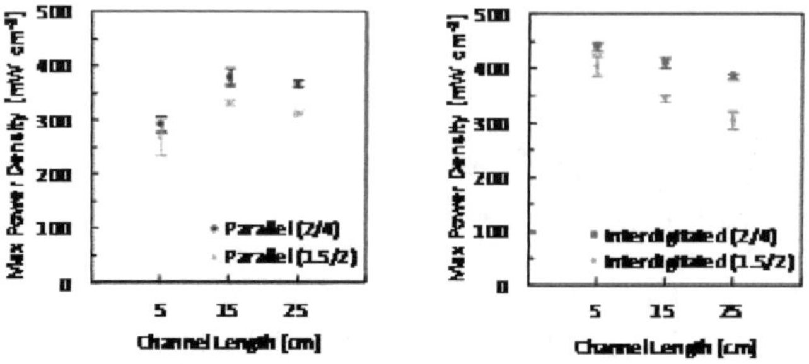

Figure 4: Summary of parallel and interdigitated max power density trends with channel length.

Figure 4: Summary of parallel and interdigitated max power density trends with channel length.

These bipolar plates can fit into a cell superstructure that has the ability to change from parallel configuration to interdigitated configuration through the use of valves and a novel manifold ? channel connection design. The valves close the inflow of gas to every other channel, and close the outflow of gas to the remaining channels. This can be done without taking apart the cell, allowing for direct in situ comparison of parallel and interdigitated flow fields.

The work of this study has already been started, with preliminary testing of a nickel plated version of the 1 x 1 mm plates already completed. These results can serve as a baseline for comparison of the gold plated set of bipolar plates. This preliminary testing has further given a better understanding of how the system behaves during operation and improvements which may be made in the new plates.

Further research will be required in the area of modeling fuel cells with electrochemical reactions to determine the effect of these channel dimensions on cell performance. Neutron radiography will need to further be performed on the plates with varying channel widths and depths to isolate the water removal characteristics of these designs. Coupling this data with performance trends should give researchers and cell industry a better understanding of design consequences.

(35b) Polymer-Based Manufacturing of Microscale Fuel Cells

Hollinger, A. S., University of Illinois
Kenis, P. J. A., University of Illinois at Urbana-Champaign
Originally presented on: 11/4/2013 8:52:00 - 9:14:00

Many microfluidic fuel cells presented to date are much too large to be integrated into portable electronic devices. In addition, most microfluidic fuel cells are unit cells which will not meet power requirements of a portable electronic device [1,2]. To address these limitations, manufacturing technologies must be developed for these fuel cells. To this end, we have designed and fabricated a polymer-based fuel cell which enables smaller, lighter structures and thinner stacks that are attractive for portable applications [3]. The polymer layers of this fuel cell were fabricated via laser ablation and sealing was achieved using a chemically-stable adhesive. This all-polymer fuel cell achieved a maximum power density of 10 mW/cm^2 at room temperature when using 1 M methanol as the fuel.

Here, we will demonstrate several polymer-based manufacturing technologies to improve the commercial viability of this fuel cell, and other microscale fuel cells. In contrast to the laser ablation used in prior work [3], we will present micro-injection molding as an alternative manufacturing technique which is much better suited for manufacturing large volumes at low costs [4]. In addition, thermal bonding of dissimilar polymer layers will be utilized as an alternative to chemical adhesion of polymer layers. For this microscale fuel cell, thermal bonding of polymer layers will greatly decrease fabrication complexity.

[1] E. Kjeang, N. Djilali, D. Sinton, Journal of Power Sources, 186 (2009) 353-369.

[2] S.A.M. Shaegh, N.T. Nguyen, S.H. Chan, International Journal of Hydrogen Energy, 36 (2011) 5675-5694.

[3] A.S. Hollinger, P.J.A. Kenis, Journal of Power Sources (2013), doi: 10.1016/j.jpowsour.2013.04.053

[4] H. Becker, C. Gartner, Analytical and Bioanalytical Chemistry, 390 (2008) 89-111.

(35c) Preparation of Pt/C Catalyst By Particle ALD, Fuel Cell Electrode Fabrication and Testing

Lubers, A. M., University of Colorado at Boulder
Anderson, K. M., University of Colorado at Boulder
Gould, T. D., University of Colorado at Boulder
Faust, M., Karlsruhe Institute of Technologie (KIT)
Seipenbusch, M., UniversitÃ¤t Karlsruhe
Weimer, A. W., University of Colorado at Boulder
Originally presented on: 11/4/2013 9:14:00 - 9:36:00

Supported platinum catalysts are in common use for hydrogen fuel cells due to their efficacy for both the hydrogen oxidation reaction (HOR) and oxygen reduction reaction (ORR). We have functionalized carbon black supports and deposited Pt nanoparticles using atomic layer deposition (ALD). In a gas-phase atomic-layer by layer process that separates the removal of organic from an organometallic, ALD was performed using trimethyl(methylcyclopentadienyl)platinum(IV) as the metal precursor reacted with either oxygen or hydrogen as the second precursor. Platinum ALD produces nanoparticles with sizes and loading depending on how many cycles are employed. The use of oxygen as a second precursor has a higher deposition rate, but particles have a larger range of sizes and are less dispersed compared to using hydrogen. Deposition on the functionalized carbon produces more uniform particle sizes and finer distribution. Electrodes were fabricated and tested. This presentation will focus on the Pt/C catalyst preparation and electrode performance using a PEM fuel cell test stand.

(35d) Understanding the Gas Diffusion Layer in PEM Fuel Cells - Experimental Characterization, Design, and Manufacturability

Morgan, J., Dept. of Chem. Eng., University of New Hampshire
Datta, R., Worcester Polytechnic Institute
Originally presented on: 11/4/2013 9:36:00 - 9:58:00

The proton exchange membrane fuel cell (PEMFC) has a significant potential in transportation, backup, and portable power applications, although there still are some remaining technical and cost challenges. A key current challenge is improving the performance while reducing the cost of the gas diffusion layer (GDL). Designing a commercial GDL, however, is far more complex than simply making a porous, sturdy, conductive layer, because of the trade-offs among performance, manufacturability, and cost. An improved understanding of its multifarious functions in the fuel cell can help attain this goal. It is, thus, important to understand how each GDL parameter affects four interrelated key outcomes: 1) performance, 2) durability, 3) manufacturability, and 4) cost. Here, we examine 11 key properties of the GDL, how they influence these outcomes, how these are measured experimentally *ex-situ*, and how they can be altered via the manufacturing process. In particular, we investigate the correlation between measured *ex-situ* measured effective diffusivity and *in-situ* performance in a PEM fuel cell, as well as the impact of the MPL structure on cell performance under both wet and dry conditions.

(35e) Hydrogen and Proton Exchange Membrane Fuel Cells for Transportation

Mathur, V. K., University of New Hampshire
Teng, X., University of New Hampshire
Originally presented on: 11/4/2013 9:58:00 - 10:20:00

Hydrogen has the potential to solve many of the energy challenges that confront the world today, such as reducing dependence on petroleum imports and decreasing pollution and green house gas emissions. There is general agreement that hydrogen could play an increasingly important role in world's energy future. Currently, we have a hydrocarbon economy but we lack the know-how to produce hydrogen from hydrocarbons or water and deliver it to consumers in a clean, affordable, safe and convenient manner as an automotive fuel or for power generation.

Hydrogen can be produced through thermal, electric or pyrolytic processes applied to fossil fuel, water or biomass. Nuclear-based systems can produce hydrogen from water using thermal or electrolytic processes but are not as efficient or cost effective as fossil fuels. The thermal production procedure that uses steam to produce hydrogen from natural gas or other light hydrocarbon is most common

The technology of fuel cell systems for electric vehicles started from the history of the Alkaline Fuel Cell (AFC) after World War II. Since then fuel cells have mostly been used in space vehicles. However, since the 1990s, there has been considerable effort to develop fuel cells for land based vehicular transportation. The Ballard Power System of Canada has successfully collaborated with several automobile companies in this area towards the development of electric vehicles. The feasibility of electric vehicles to operate on fuel cell power is a fact. The choice of fuel cell is still open. The prototypes of fuel-cell operated small appliances for public use may be in the market within two to three years. However, fuel-cell driven automobiles may not be available commercially for another decade.

A number of fuel cell vehicles are being developed and tested by several car manufacturers including Ford Motor Co (USA), General Motors Corp (USA), DaimlerChrysler AG (Germany), Honda Motors Co (Japan), Toyota Motor Co (Japan), Renault- Nissan (France/Japan) and PSA Peugeot Citroen (France). In the USA, as in other countries, the factors adversely affecting hydrogen's potential as a major energy source are our inability to build and sustain national consensus on energy policy priorities; the lack of a hydrogen infrastructure and substantial cost of building one; the high cost of hydrogen production, storage and conversion devices such as fuel cells; and hydrogen safety issues.

The reliance upon the oil, gas and coal for all our energy needs has resulted in the severe depletion of these natural resources and caused immense environmental pollution problems. Other alternative energy technologies such as solar, wind, hydroelectric power can provide power for many applications but are not suitable to be used for transportation vehicles, military applications, and the long-term needs of future electronics. One option is to use hydrogen which has high energy density and is non-polluting. It can be used directly in internal combustion engines or in fuel cells for generating electricity. For transportation applications, fuel cell-operated vehicles can be supplied hydrogen generated at the fueling stations or transported from production plants. There is a need for the development of an infrastructure for hydrogen fueling stations which is practically nonexistent to this date. It is difficult to predict how soon the transition to hydrogen economy will occur. Cost of hydrogen production, transportation and fuel cell operated vehicles has to come down before future generation can enjoy the benefits of electric vehicles. Investment in R&D will help us move towards this goal.

(35f) Surface Dealloyed Pt Nanoparticles Supported On Carbon Nanotubes: Facile Synthesis and Promising Applications for Direct Crude Glycerol Anion-Exchange Membrane Fuel Cell

Qi, J., Michigan Technological University
Xin, L., Michigan Technological University
Chadderdon, D., Michigan Technological University
Qiu, Y., Michigan Technological University
Jiang, Y., Michigan Technological University
Sun, K., University of Michigan
He, H., Michigan Technological University
Zhang, Z., Michigan Technological University
Asthana, A.,
Liang, C., Dalian University of Technology
Li, W., Michigan Technological University
Originally presented on: 11/4/2013 10:20:00 - 10:42:00

Surface dealloyed PtCo nanoparticles supported on carbon nanotube: facile synthesis and promising application for anion exchange membrane direct crude glycerol fuel cell

Ji Qi[1], Le Xin[1], David Chadderdon[1], Yang Qiu[1], Yibo Jiang[2], Kai Sun[3], Haiying He[4], Zhiyong Zhang[1], Changhai Liang[5], Wenzhen Li[1]

[1]*Department of Chemical Engineering, Michigan Technological University*

[2]*Department of Civil & Environmental Engineering, Michigan Technological University*

[3]*Department of Materials Science and Engineering, University of Michigan*

[4]*Department of Physics, Michigan Technological University*

[5]*School of Chemical Engineering, Dalian University of Technology*

[6]*Department of Materials Science and Engineering, Michigan Technological University*

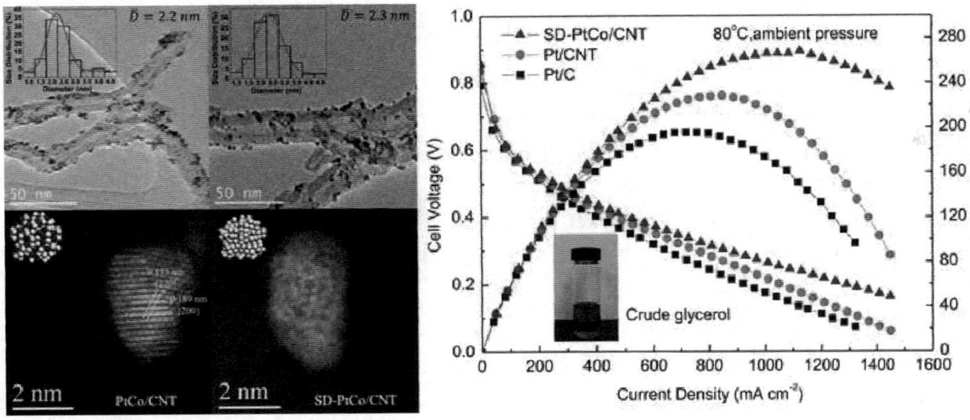

Due to the fast growth of global energy needs and quickly diminishing of fossil fuel resources, people are forced to seek reliable, high performance, cost-effective and environmentally-beneficial renewable energy sources. Anion exchange membrane based direct alcohol fuel cells have recently attract enormous attention as a potential solution to alleviate the current energy issues.

Exploring crude glycerol as fuel for direct alcohol fuel cells not only provides a promising solution to using excessive biodiesel byproduct, but also opens a new avenue towards development of low-cost alcohol fuel cells. In the present study, surface dealloyed PtCo nanoparticles supported on carbon nanotube (SD-PtCo/CNT) were prepared by *ex situ* method and used for crude glycerol oxidation for the first time. SD-PtCo/CNT anode catalyst based AEMFC with a 0.5 mg_{Pt} cm^{-2} achieved peak power densities of 268.5 mW cm^{-2} (crude glycerol/O_2) and 284.6 mW cm^{-2} (high purity glycerol/O_2) at 80 °C and ambient pressure, which are close to the published result of direct high purity glycerol solid oxide fuel cell operated at high temperature of 800 °C (327 mW cm^{-2}), and are higher than all other published performances of direct high purity glycerol microbial fuel cell and anion exchange membrane fuel cell. This work successfully developed a high output power direct alcohol fuel cells with biorenewable, environmentally-friendly fuel and dealloy technique prepared catalyst, which will substantially impact future catalyst design and fuel cell development.

(35g) Mechanism of Imidazolium Groups Degradation in Alkaline Anion Exchange Membrane

Wang, W., Tsinghua University
Xie, X., Tsinghua University
Lv, Y., Beijing University of Chemical Technology
Originally presented on: 11/4/2013 10:42:00 - 11:04:00

Recently, alkaline electrolyte membrane fuel cells (AEMFCs) have been attracting growing interest. By changing acidic medium into a basic one, the cathode oxygen reduction over-potential can be greatly reduced, resulting in higher efficiency, and the catalysts are more stable in alkaline environment. Moreover, the costs of the fuel cells can be dramatically reduced because AEMFCs can work with non-noble metal catalysts such as nickel and silver.

Since imidazolium cations tend to be attacked by hydroxide, the membrane degradation occurs over time. Those greatly limit the development of alkaline electrolyte membrane fuel cells. Imidazolium groups as cationic head groups are prone to degradation in alkaline conditions through a ring-opening mechanism. In the mechanism, a hydroxide ion performs a nucleophilic attack on the imidazolium ring at α carbon (α carbon with reference to both nitrogen atoms). In this study, density functional theory calculations were used to research the degradation pathways of imidazolium cations to discuss the relative durability of cation bounded by alkyl and aromatic groups. Detailed structure changes of imidazolium ring were explored by means of quantum chemical calculations (QCCs) using a model side chain terminal of dimethylimidazolium (DMIM) cation. The reaction energy barrier of the degradation was obtained from our reaction path calculations. The effect of the medium and the water content was also discussed on the calculated reaction barriers. Good solvation of the head-groups and hydroxide ions is helpful to the chemical stability of alkaline membranes.

(50a) Crosslinked Anion Exchange Membranes for Redox Flow Batteries

Jung, M. S., Illinois Institute of Technology
Parrondo, J., Illinois Institute of Technology
Arges, C., Illinois Institute of Technology
Ramani, V., Illinois Institute of Technology
Originally presented on: 11/4/2013 8:30:00 - 8:55:00

All-vanadium redox flow batteries (VRFB) have been very popular as they use the same electrochemical species in both electrode chambers and therefore they do not suffer from electrolyte cross contamination. A core component of any RFB is the electrolyte membrane (PEM or AEM) used to separate the positive and negative electrolytes. The key requirements for the separator are good chemical and electrochemical stability, good mechanical stability, high ionic conductivity, and low permeability for the vanadium ions. Nafion® membranes have been the most popular separators in RFBs because of their excellent oxidative stability and high proton conductivity, however, they exhibit high vanadium crossover. Anion exchange membranes (AEMs) are ideal candidates as RFB separators as they have much lower vanadium ion crossover than proton exchange membranes (10-100 times less) due to Donnan exclusion. Moreover, the reduced penetration of VO_2^+ into the AEM minimizes chemical degradation.

We selected poly (2,6-dimethyl 1,4-phenylene) oxide (PPO) based AEMs as battery separators as they possess good conductivity, low water uptake and good mechanical stability. PPO is commercially available, relatively inexpensive, and can be readily modified by attaching bromine to the backbone to induce a phase-separated morphology. PPO can be easily functionalized into an AEM by bromination followed by reaction with a tertiary amine. In this talk, we will discuss the effect of polymer crosslinking on the conductivity, water uptake, vanadium (IV) permeability, ultimate tensile stress, glass transition temperature and storage modulus of PPO- based AEMs. Polymer crosslinking has been performed by the reaction of brominated PPO (Br-PPO) with tetramethylhexanediamine (TMHDA) in liquid phase. Non crosslinked AEMs have been prepared by reaction of the Br-PPO with quinuclidine ($PPO-ABCO^+$) and trimethylamine ($PPO-TMA^+$). Nuclear Magnetic Resonance techniques [^1H-NMR, ^{13}C-NMR, correlation spectroscopy (COSY), and ^1H-^{13}C heteronuclear multiple quantum correlation (HMQC) spectroscopy] have been used to investigate the polymer structure before and after exposure to VO_2^+ solution (over time) and to identify any possible degradation products and mechanisms. Crosslinked membranes will be tested (and compared against Nafion® and non crosslinked AEMs) in a vanadium redox flow battery to evaluate its performance and durability. Postmortem analysis of the membranes via NMR, conductivity/IEC measurements, and mechanical testing will be performed to unearth key degradation modes.

(50b) Challenges to Implementation of High-Rate Energy Storage in Military Applications

Heinzel, J., Naval Sea Systems Command - Philadelphia
Ostanek, J.,
Hoffman, D., Office of Naval Research
Peek, I., Naval Sea Systems Command - Philadelphia
Originally presented on: 11/4/2013 8:55:00 - 9:20:00

Future military platforms will require more extensive use of electronic power systems to achieve required performance levels. Energy storage is becoming a key enabler for power systems operating with high levels of transient power requirements and supporting fuel-efficiency operations while maintaining capability. High density energy storage which operates continuously at high charge and discharge rates to support these loads under the combination of finite space, high power and energy requirements, stringent safety characteristics and wide upper temperature ranges drive the requirement for advancement in a multiphysical mix of electrical, chemical, thermal, safety and controls elements, which must be considered and have development performed to enable widespread implementation. These will be discussed from the perspective of requirements and considerations for system design and engineering for applications associated with high rate support of transient loads, power quality and optimization of generator and electrical architecture dynamics.

(50c) Novel Pt/TixM1-xO2 Nanocatalysts for Fuel Cells

Hwang, B. J., National Taiwan University of Science and Technology
Ho, V. T. T.,
Pan, C. J., National Taiwan University of Science and Technology
Su, W. N.,
Originally presented on: 11/4/2013 9:20:00 - 9:45:00

This work presents a new approach by exploring robust non-carbon $Ti_xM_{1-x}O_2$ as a novel functionalized co-catalytic support for Pt. This new approach is based on the novel nanostructure of the $Ti_xM_{1-x}O_2$ supports which holds strong metal-support interactions (SMSI) between $Ti_xM_{1-x}O_2$ and Pt that can not only modify surface electronic structure of Pt, evidenced from a shift in the d-band centre of the surface Pt atoms but also provide dual reaction sites for methanol oxidation. Furthermore, another benefit of $Ti_xM_{1-x}O_2$ is the extremely high stability of Pt/ v during potential cycling, which is attributable to the SMSI between Pt and $Ti_xM_{1-x}O_2$. Interestingly, $Ti_xM_{1-x}O_2$ can be fabricated as a much thinner catalyst layer, resulting in significantly improved the mass transport kinetics and performance of the resulting membrane-electrode-assembly (MEA).The new approach presented in this work opens a reliable path to the discovery of advanced concepts that may lead to the design of new catalyst materials that can replace the traditional catalytic structures and motivate further research in this field.

(50d) A Tunable Hierarchical Porous Carbon Structure for Studying Electrochemical Energy Conversion

Chan, K. Y., The University of Hong Kong

Yang, C.,

Originally presented on: 11/4/2013 9:45:00 - 10:10:00

A Tunable Hierarchical Porous Carbon Structure for Studying Electrochemical Energy Conversion

Chunzhen YANG and Kwong-Yu CHAN*, Department of Chemistry, The University of Hong Kong, Pokfulam Road, Hong Kong

Abstract

A well-defined porous carbon structure with tunable parameters provides a platform for systematic study of transport and confinement effects on electrochemical conversion. Synthesized via silica template techniques, the hierarchical structure has a hollow core within a mesoporous shell (HCMS) with uniform pore size, shell thickness, hollow core diameter, as shown in Fig. 1.

By varying the synthesis parameters, the structural parameters can be tuned to study

1) combined effects of surface area, mesopore size, macropore size, and hierarchy of porosity on ionic transport and adsorption in electrochemical capacitance behavior;

2) how distribution of platinum or other electrocatalyst nanoparticles in the porous structure affect catalytic faradaic reactions;

3) how distribution of mixed metal nanoparticles such as PtRu and varying of Pt and Ru content in the porous structure affects faradaic reactions such as methanol oxidation; and

4) the effectiveness of a mathematic model in analyzing electrochemical conversion in porous structures.

Electrochemical capacitance is studied with a series with meso-shell thickness stepwise increased from 0, 25, 50 to 100 nm while keeping an identical 330 nm hollow core and mesopore of 3.9 nm.[1] A thicker shell has a higher surface area with a proportional increase of electrochemical capacitance which however, can only be fully realized at low scan rates/currents. At high currents, ionic transport limits the electrochemical capacitance of a thick mesoporous shell. Electrochemical impedance spectra (EIS) obtained on the family of HCMS carbon structures were fitted to equivalence circuits described by models of network structures.[2] The results reveal the need to match the AC frequency with the characteristic time constant of a structure for optimum active and reactive power. For fuel-cell electrode conversion, the distribution of Pt and PtRu nanoparticles in a mesoporous carbon can vary in different synthetic protocol. Uniform Pt particle distribution can be achieved with a CPDP method [3] as shown in Fig. 2. Different electrocatalysts with varying Pt distribution can be compared for activity and utilization in methanol oxidation. The particles dispersed into a thicker shell are more stable and give better performance over time, as shown in voltage cycling and extended methanol oxidation.

SHAPE * MERGEFORMAT

Fig. 1. A hollow core mesoporous shell (HCMS) carbon structure with three varying geometrical parameters: pore diameter (d_p), shell thickness (S), and diameter of core (C).

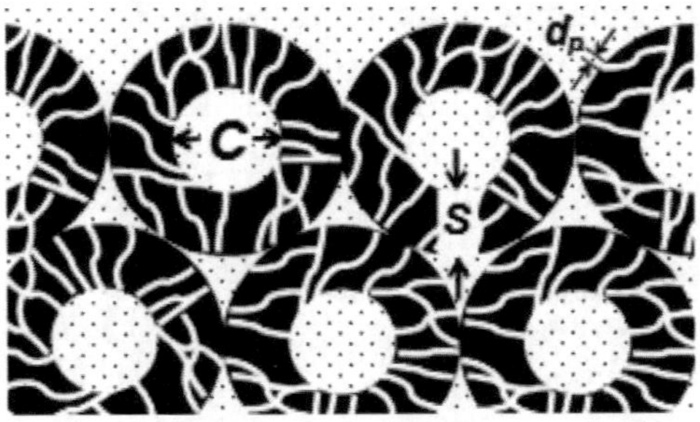

Fig. 1. A hollow core mesoporous shell (HCMS) carbon structure with three varying geometrical parameters: pore diameter (d_p), shell thickness (S), and diameter of core (C).

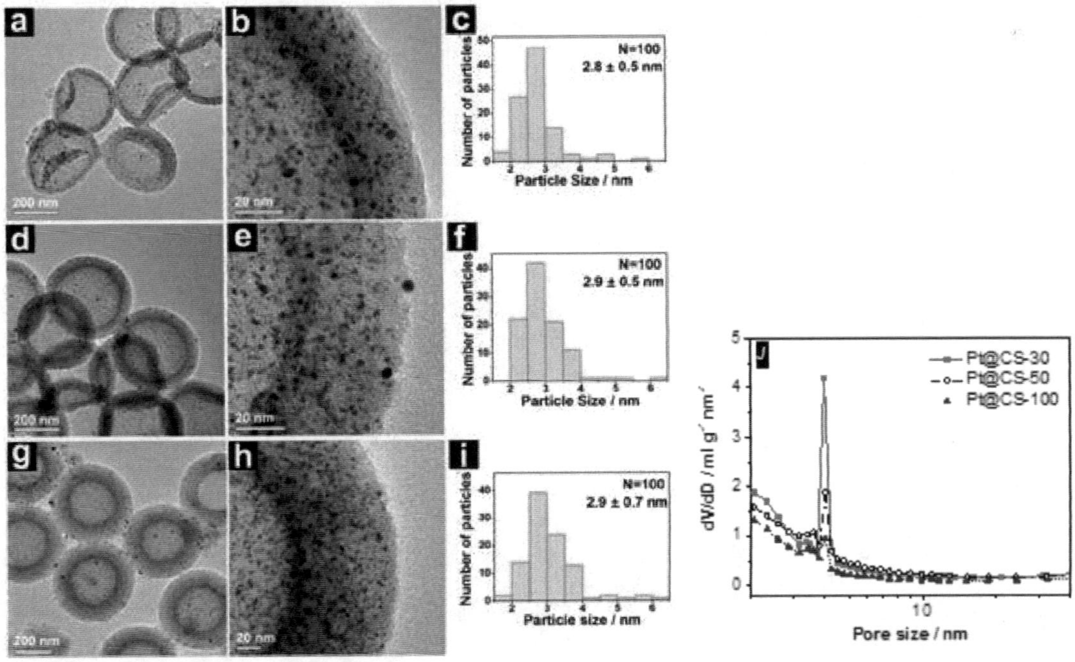

SHAPE * MERGEFORMAT

Fig. 2 TEM images of Pt nanoparticles loaded in mesoporous carbons of same hollow core diameter, *C* = 330 nm but different Shell thickness (a)-(c) Pt@CS-30, *S*=30 nm ; (d)-(f) Pt@CS-50, *S*=50 nm ; (g)-(i) Pt@CS-100, *S*=100 nm; (j) pore size distribution of the porous structures loaded with Pt.

Fig. 2 TEM images of Pt nanoparticles loaded in mesoporous carbons of same hollow core diameter, *C*=330 nm but different Shell thickness (a)-(c) Pt@CS-30, *S*=30 nm ; (d)-(f) Pt@CS-50, *S*=50 nm ; (g)-(i) Pt@CS-100, *S*=100 nm; (j) pore size distribution of the porous structures loaded with Pt.

References:

[1]. F. Li, M. Morris and K.Y. Chan*, ?Electrochemical Capacitance and Ionic Transport in the Mesoporous Shell of a Hierarchical Porous Core-Shell Carbon Structure?, *J. Mater. Chem.* 21 (2011) 8880-8886.

[2] Chunzhen Yang, Chi-Ying Vanessa Li, Fujun Li, Kwong-Yu Chan*, ?Complex Impedance with Transmission Line Model and Complex Capacitance Analysis of Ion Transport and Accumulation in Hierarchical Core-Shell Porous Carbons?, *J. Electrochem. Soc.* (2013) to appear.

[3] F. Li, K.Y. Chan*, and H. Yung, ?Carbonization over PFA-Protected Dispersed Platinum: An Effective Route to Synthesize High Performance Mesoporous-Carbon Supported Platinum?, *J. Mater. Chem.* 21 (2011) 12139-12144.

(50e) High-Performance Electrodes for Hydroxide Exchange Membrane Fuel Cells

Kaspar, R. B., University of Delaware
Letterio, M., University of Delaware
Jensen, K., University of Delaware
Gu, S., University of Delaware
Yan, Y., University of Delaware
Originally presented on: 11/4/2013 10:10:00 - 10:35:00

Hydroxide exchange membrane fuel cells (HEMFCs) are an attractive alternative to conventional proton exchange membrane fuel cells (PEMFCs) because they are compatible with non-precious-metal catalysts. However, HEMFC performance is lower than the PEMFC benchmark, in part because HEMFCs continue to borrow electrode architectures that were originally designed for PEMFCs. In PEMFCs, water is produced at the cathode, so hydrophobic gas diffusion layers (GDLs) are employed to control flooding. But hydrophobic GDLs may not be necessary for HEMFC cathodes, in which water is consumed instead of produced, reversing the direction of water transport in the cell. This work demonstrates that the performance and durability of membrane-electrode assemblies (MEAs) can be improved by tailoring electrode design to reflect the unique kinetics and water transport behavior of HEMFCs.

(50f) Capacitive Mixing in Porous Electrodes for Energy Production

Gabitto, J. F., PVAMU
Tsouris, C., Oak Ridge National Laboratory
Originally presented on: 11/4/2013 10:35:00 - 11:00:00

Energy obtained by salinity-gradients (blue energy) is a promising renewable energy source. Mixing of a high-concentration saline solution, such as seawater, with fresh water (river water) to produce a brackish solution dissipates more than 2.2 MJ of free energy per m^3 of fresh water. The charging-discharging cycle inside the pores of the solid electrodes is strongly influenced by the pore size distribution. Macroporous materials behave like standard plate capacitors, while mesoporous materials behave like cylindrical capacitors. The goal of this project was to optimize the recovery efficiency of free energy change by using tailored-design carbon electrodes. Morphology, pore size, and carbon electrode composition were chosen to optimize the charge-discharge steps of the reversible cycle. The physicochemical processes inside the electrode and the corresponding effects on the reversible energy cycle were investigated. A proposed energy cycle comprises a charging step, a discharging step and two switching steps that replace a saltwater solution inside the pores by a fresh water solution and vice versa. Equations derived using the volume averaging method have been employed to simulate the charging-discharging cycle in homogeneous porous materials of a macropore size distribution. Ionic transport by diffusion and electrical double layer expansion processes were considered to simulate the two alternating steps. Improved energy production was achieved by optimizing the duration of each step and variation of the operating variables.

(65a) Templated Ordered Mesoporou Materials for Supercapacitors and Lithium Ion Batteries

Lee, J., Pohang University of Science and Engineering

Originally presented on: 11/4/2013 8:30:00 - 9:00:00

For high performance supercapacitor and Li-ion battery electrodes, various mesoporous metal oxide materials have been synthesized by employing lab-synthesized block copolymers as a structure-directing agent or ordered mesoporous silica materials as hard templates. An ordered mesoporous WO_{3-X} material with high electrical conductivity was employed for use as a supercapacitor electrode in 2.0 M H_2SO_4. The electrode made of mesoporous WO_{3-X} exhibited a high rate capability and an excellent capacitance (366 Fcm^{-2}, 639 Fcm^{-3}), which were attributed to the large ordered mesopores, high electrical conductivity, and high material density. Fe_3O_4 nanocrystals confined in mesocellular carbon foam (MSU-F-C) are synthesized by 'Host-Guest' approach and tested as anode for lithium-ion batteries (LIBs). 13-27 nm magnetite Fe3O4 nanocrystals are formed inside ~30 nm pores of mesocellular carbon matrix. Fe_3O_4/MSU-F-C (45 wt% Fe_3O_4) exhibited 1007 mA h g^{-1} of first charge (Li de-insertion) capacity at 0.1 A g^{-1} and 111% of capacity retention at 150th cycle, and retains 37% capacity at 7 A g^{-1} (~7 C rate). For the first time, we synthesized mesostructured LTO/carbon composite (Lithium titanate/carbon) with large (> 15 nm) and uniform pores for high performance LIB anodes via a block copolymer assembly. Sn-mesostructured carbon nanocomposites have been synthesized by simple self-assembly of block copolymers with Tin and carbon precursors.

(65b) Reduced Graphene Oxide Wrapped FeS Nanocomposite for Lithium-Ion Battery Anode With Improved Performance

Fei, L., New Mexico State University
Luo, H., New Mexico State University
Originally presented on: 11/4/2013 9:00:00 - 9:15:00

The fast growing business of portable electronics (e.g. computers, mobile phones, cameras) has stimulated the rapid development of lithium-ion batteries (LIBs) due to their high energy density, flexible and lightweight design, and longer lifespan than conventional battery technologies. Graphite is the most popular commercial anode. However, its low theoretical capacity (372 mAh/g) renders it hard to satisfy the increasing market demands. Recently, metal sulfides with high theoretical capacity are widely studied as another promising class of anode materials with excellent electrochemical performance.

Here, we report FeS based nanocomposite as anode for lithium-ion batteries. FeS nanoparticles wrapped in reduced graphene oxide (RGO) are produced via a facile direct-precipitation approach. The resulting nanocomposite FeS@RGO structure has better lithium ion storage properties exceeding those of FeS prepared without RGO sheets. The nanocomposite delivered an initial discharge capacity of 1357 mAh/g and a charge capacity of 1116 mAh/g, giving the first cycle coulombic efficiency of 82 %. The composite shows a very stable cyclic performance with its capacity remaining as high as 978 mAh/g even after 40 cycles at a current density of 100 mA/g. This value is almost three times the capacity of commercial graphite anode, and is comparable to those reported in literature. At current density of 300 mA/g, the composite delivered discharge capacity of 927, 690, and 618 mAh/g in the first, 10th, and 30th discharge cycles, respectively; while FeS suffered rapid capacity decline, only delivering 885, 518, and 382 mAh/g in the first, 10th, and 30th discharge cycles. Electrochemical impedance demonstrate that FeS@RGO has a lower electrochemical resistance than that of FeS. The enhanced electrochemical performance is attributed to the robust sheet-wrapped structure and the synergetic effects between FeS and RGO sheets, such as increased conductivity, shortened lithium ion diffusion path, and the effective prevention of polysulfide dissolution.

(65c) Alkanethiol-Passivated Germanium Nanowires As High-Performance Anode Materials for Lithium-Ion Batteries

Tuan, H. Y., National Tsing Hua University
Yuan, F. W., National Tsing Hua University
Originally presented on: 11/4/2013 9:15:00 - 9:30:00

Rechargeable lithium-ion batteries have been widely used in portable electronic devices such as cell phones, laptop computers, and digital cameras, and have been considered as an electrochemical energy storage system for powering hybrid electric vehicles (HEV) and zero-emission vehicles (ZEV). It is necessary to develop lithium-ion batteries having a higher energy density and a higher power density to meet the increasing energy demand. Germanium (Ge) has emerged as a promising anode materials due to its high theoretical capacity of 1384 mAh/g - around four times greater than commercial graphite anode (372 mAh/g). However, the practical usage of Ge as an anode material is hindered by dramatic volume changes (~300%) during the insertion/extraction process of lithium ions, causing crack and pulverization, and loss of electrode contact.

Ge nanowires can effectively accommodate the volume changes, tolerate relaxed mechanical strain, provide channels for efficient electron transport, and alleviate pulverization. Although direct incorporation of Ge nanowires into slurry coating process shows improved electrochemical performance relative to microparticles, the anode performance decayed rapidly within a few tens of cycles. In this talk, we demonstrate that dodecanethiol monolayer passivation on Ge nanowire surface can significantly enhance their anode performance in lithium-ion batteries. The dodecanethiol-passivated Ge nanowires exhibit an excellent electrochemical performance with a reversible specific capacity of 1130 mAh/g at 0.1 C rate over 100 cycles, which is above 80% of the Ge theoretical capacity. The functionalized Ge nanowires show high-rate capability having charge and discharge capacities of ~555 mAh/g at rates as high as 11 C. An aluminum pouch type lithium cell using a $LiFePO_4$ cathode was also assembled to provide larger current (~30 mA) for uses on light-emitting-diodes (LEDs) and audio devices. Organic monolayer-coated Ge nanowires represent promising Ge-C anodes with very low carbon content (~2-3 wt %) for high capacity, high-rate lithium-ion batteries and are readily compatible with commercial slurry-coating process for cell fabrication.

(65d) A Facile Microwave-Assisted Route to Co(OH)2 and Co3O4 Nanosheet for Li-Ion Battery

Chen, G., New Mexico State University
Luo, H., New Mexico State University
Originally presented on: 11/4/2013 9:30:00 - 9:45:00

Cobalt hydroxide ($Co(OH)_2$), as one of the most important functional inorganic materials, can be used in the field of catalysis and electrochemical electrodes. The previous report revealed that $Co(OH)_2$ is a promising alternative conversion-based anode materials for lithium-ion batteries (LIBs) due to its high capacity. Unfortunately, like other conversion-based anode materials, the implementation in LIBs is greatly hampered by its poor cycling performance. One of the most promising strategies to tackle this obstacle is to make $Co(OH)_2$ into sheet-like structure to improve the structural stability in the electrochemical reactions. Compared with the bulk $Co(OH)_2$, $Co(OH)_2$nanosheet displays obvious advantages due to large surface area, highly oriented growth direction, tunable interlayer space for the de/insertion of lithium ions.

In this work, a facile microwave-assisted synthetic route has been successfully developed for preparing hexagonal $Co(OH)_2$ nanosheets with average width of 2 μm and thickness of 100 nm. $Co(OH)_2$ can further convert to porous Co_3O_4 nanosheets *via* thermal decomposition. Their phases, structures and morphologies were investigated by X-ray diffraction (XRD), atomic force microscopy (AFM), and transmission electron microscopy (TEM). Both $Co(OH)_2$ and Co_3O_4 nanosheets can serve as potential candidates for anodes of Li-ion battery. The electrochemical study revealed that $Co(OH)_2$ and Co_3O_4 nanosheets delivered a reversible capacity of 600 and 700 mAh g^{-1} after 40 cycles, respectively. Cyclic voltammetry (CV) curves also confirmed the relative stability of the as-synthesized electrode materials. This effective microwave-assisted route may be a promising approach for preparing other transition metal hydroxide/oxides for energy applications.

(65e) Li Adsorption and Intercalation in Graphene, Graphite and Carbon Onions By Reaxff Reactive Force Field

Raju, M., The Pennsylvania State University
Ganesh, P., Oak Ridge National Laboratory
Kent, P. R. C., Oak Ridge National Laboratory
van Duin, A. C. T., Pennsylvania State University
Originally presented on: 11/4/2013 9:45:00 - 10:00:00

A ReaxFF reactive force field has been developed to describe Li/C systems. The ReaxFF force-field parameters for Li-C systems were optimized against a density-functional theory (DFT) (with van der Waals corrections) based training set, containing a collection of data (energies, geometries, charges etc.) related to adsorption and diffusion of Li on graphene and in graphite. ReaxFF reproduces accurately the adsorption energies for Li on hollow, bridge and top sites on pristine graphene, formation energy of stage I compounds in graphite and intralayer Li migration barriers in graphite at separations of 3.35 Å and 3.7 Å. To validate this force field we have compared it against DFT calculations for Li diffusion in the direction perpendicular to the graphene sheet for graphene with single and double-vacancy. We performed grand canonical Monte Carlo simulations to obtain ground states for Li intercalation in graphite and find that ReaxFF describes the in-plane Li ordering and interlayer separations for stage I and II compounds in good agreement with previous DFT studies and experiment. The voltage profile for the Li-graphite system as a function of Li concentration obtained by ReaxFF matches experiment and DFT data. We study Li intercalation in presence of vacancy defects in graphite and find that the presence of vacancy defects increase the ratio of Li/C and shifts the voltage profile downwards, both in proportion to the number of vacancy defects. Li intercalation and diffusion kinetics in Carbon Onions was investigated for potential applications in batteries and electrochemical capacitors and we report a different loading behavior/voltage profile for these Carbon Onions as compared to graphitic systems. The Voltage vs Li concentration for Carbon Onions has a 'saw tooth' shape whereas the voltage profile for graphitic systems consists of plateaus.

(65f) Silicon-Rich Carbon Nanofibers From Water-Based Spinning to Control Combination of Silicon and Carbon for Li-Ion Battery

Joo, Y. L., Cornell University
Kim, Y. S., Cornell University
Kim, K. W., Cornell University
Cho, D.,
Hansen, N. S., Cornell University
Originally presented on: 11/4/2013 10:00:00 - 10:15:00

Silicon-rich carbon nanofibers from water-based spinning to control combination of silicon and carbon for Li-ion battery

Yong Seok Kim, Kyoung Woo Kim, Daehwan Cho, Nathaniel S. Hansen, and Yong Lak Joo*

School of Chemical and Biomolecular Engineering, Cornell University, Ithaca 14853, USA

ABSTRACT: Carbon (C) nanofibers with high silicon (Si) loading (> 65 wt%) were fabricated from water–based electrospinning of aqueous polyvinyl alcohol/Si nanoparticle solution for Li–ion battery anode applications. Our Si-rich C (SRC) nanofibers showed much facilitated charge transport features and increased activities because of continuous one–dimensional (1–D) carbon backbone structure, leading to superior battery performance to that of bare silicon nanoparticles. The presence of carbon as 1–D nanofibers could not only mitigate volume expansion of silicon but also avoid formation unstable solid electrolyte interface onto the surface of silicon. Our study on the optimum combination of carbon and silicon in the nanofibers for their electrochemical properties and battery performance revealed that the SRC nanofibers with 72.8 wt% Si content (27.2 wt% C) exhibits an adequate balance between high energy capacity of Si nanoparticles and dimensional stability with effective charge transport by carbon nanofibers, leading to an outstanding cycle life to maintain 1076 mAh/g normalized by total electrode mass and high Coulombic efficiency of about 99% for fifty cycles. Such scalable Si-rich C nanofibers made from water-based electrospinning using polyvinyl alcohol should be greatly contributed to a cost–effective development of high performance negative battery electrodes.

(65g) Electrochemical Properties of Carbon-Coated SnO2-Graphene Nanocomposites for Rechargeable Lithium-Ion Batteries

Ara, M., Wayne State University
Salley, D. S., Wayne State University
Ng, S. K. Y., Wayne State University
Originally presented on: 11/4/2013 10:15:00 - 10:30:00

SnO_2 nanoparticles were carbon-coated using two different methods and then dispersed on the reduced graphene oxide nanosheets prepared from modified Hummer's method. Carbon-coating was performed by a solvothermal method from glucose precursor material or by carbonization in Polyacrylonitrile (PAN) based carbon fibers.The later was prepared by dispersing SnO_2 nanoparticles in N-Methyl-2-pyrrolidone and then the dried powder was stabilized in air at 270°C for 30 minutes followed by the heat treatment at 900°C for 2 hours. The coated nanocomposites were characterized by X-ray diffraction, scanning electron microscopy (SEM), transmission electron microscopy (TEM), and Raman spectroscopy. From TEM observations, it was found that the SnO_2 nanoparticles were 2-6 nm in diameter and were uniformly distributed on the reduced graphene oxide nanosheets. Electrochemical properties were evaluated by galvanostatic charge-discharge cycling. Carbon-coated SnO_2-reduced graphene oxide nanocomposites showed higher capacity and more stable electrochemical performances than non-coated SnO_2-reduced graphene oxide nanocomposites. Carbon coating enhances the conductivity of the electrode and it reduces the stress developed by the volume changes of the SnO_2 nanoparticles during charge/discharge cycles. The synergistic effect of the amorphous carbon, SnO_2 nanoparticles and reduced graphene oxide nanosheets is also responsible for the superior specific capacity of carbon-coated anode material. Excellent rate capability and cyclic performances were also demonstrated with higher current densities of 500 and 1000 mA/g.

(65h) Multilayer Hybrid Electrodes of TiO2/Mwnt for Electrochemical Applications

Hyder, M. N., Massachusetts Institute of Technology
Kavian, R.,
Shao-Horn, Y., Massachusetts Institute of Technology
Hammond, P. T., Massachusetts Institute of Technology
Originally presented on: 11/4/2013 10:30:00 - 10:45:00

Rational design of next-generation electrochemical devices requires novel electrode materials and architectures to achieve high-energy, high-power density to bridge the gap between battery and capacitors. Conventional electrode fabrication processes such as doctor blading allows limited control over the electrode thickness and structure at the nanoscale, in addition to the incorporation of insulating binder materials that limits device performance. To address these challenges, we have engineered a route for the synthesis of highly stable, sub-8nm TiO_2 nanoparticles that were incorporated with acid-functionalized multi-walled carbon nanotubes (MWNT). Using electrostatic layer-by-layer nanofabrication, we assembled binder-free hybrid electrode of TiO_2/MWNT to obtain a synergistic effect from high electronic conductivity and excellent charge storage capacity. These thin film electrodes with highly controllable thicknesses shows well developed mesopores with finely dispersed, non-agglomerated TiO2 nanoparticles on MWNTs. Electrochemical measurements show high charge storage capacity (>150 mAh/$g_{electrode}$ at 0.1 A/$g_{electrode}$) with excellent cycling stability during charging and discharging up to 200 cycles indicating their promise as thin-film negative electrodes for future Li storage applications.https://aiche.confex.com/aiche/2013/t5/papers/index.cgi?username=341047&password=343273

(65i) Facile Preparation of Mesoporous α-Fe2O3 With Unique Cocoon-Like Hollow Nanostructure for Reversible Lithium Ion Storage

Zhu, J., Wayne State University
Ng, K. Y. S., Wayne State University
Deng, D., Wayne State University
Originally presented on: 11/4/2013 10:45:00 - 11:00:00

After 20 years of development since the first commercialization of rechargeable lithium ion batteries by Sony in 1991, carbon is still exclusively used as anode materials for lithium ion batteries. However, the theoretical capacity (372mAh g^{-1}) of graphite is almost achieved now and it becomes a bottleneck to further increase lithium-ion storage capacity using carbon. Carbon-alternative materials with higher capacity are currently being extensively explored to meet the increasing demand for energy in mobile electronic devices, electric vehicles and energy storage in grid. Transition metal oxides receive increasing attention as carbon-alternatives because of their much higher theoretical capacity. For example, α-Fe_2O_3 with a theoretical capacity of 1007 mAh g^{-1} is a promising candidate. Compared to other transition metal oxides, α-Fe_2O_3 is outstanding in terms of low cost, abundance, ease of fabrication, and environmental benignity. However, the poor cyclability caused by volume change during the insertion /extraction of Li ions and poor conductivity is the main challenge. Our strategy is to design and tailor α-Fe_2O_3 at nanoscale with structure control to improve their electrochemical performance. In this presentation, the strategy to facilely prepare mesoporous α-Fe2O3 with unique cocoon-like hollow nanostructure as anode materials will be reported. The formation mechanism is revealed by extensive characterization by FESEM, TEM and XRD for samples prepared under different controlled conditions. The improved electrochemical performance in reversible lithium ion storage will be discussed.

(103a) Towards the Development of Economic, High Specific Energy Metal-Air Batteries in Vehicle Propulsion Applications

Chen, L. D., SUNCAT Center for Interface Science and Catalysis
Luntz, A., IBM Almaden
Norskov, J. K., Technical University of Denmark
Originally presented on: 11/4/2013 12:30:00 - 12:52:00

Since the advent of the automobile, earth's petroleum supply has been undergoing increasingly rapid depletion as private vehicles become more commonplace. Since the majority of petroleum is used for vehicle propulsion, the scientific community recognizes the imminent need to obtain the widespread implementation of an alternative energy carrier for transportation. The most obvious alternative would be the electric vehicle (EV). However, the range of EVs is still far from adequate, and this is a result of the low specific energy inherent to current lithium-ion systems. Many other caveats exist, including the high cost of lithium and in particular the large number of safety concerns associated with the system. One solution to the lithium-ion system's problems — the lithium-*air* system — was first proposed in the 1970s and is regaining its popularity as can be seen by the rise of relevant publications in the last few years. The cathode in the lithium-air battery is just air from the atmosphere (i.e. there is essentially an *infinite* supply of the cathode material), and the support can be porous carbon with a high surface area, leading to a tremendous reduction in the weight of battery. The anode is pure bulk metal instead of intercalated metal atoms, which again reduces the weight of the battery. Moreover, since one of the reactive species is not stored in the battery but instead uptaken when necessary, there is much less concern over its safety.

Taking it a step further, other metals could in principle be substituted for lithium in order to reduce the costs of production. For example, aqueous magnesium- and aluminum-air batteries theoretically possess nearly as high specific energies as lithium-air and are much more abundant, but have not been well researched or developed yet. Most studies in current technical literature are simply trying to build batteries without conducting fundamental research that probe the basic mechanisms and performance-limiting parasitic reactions. It would be greatly beneficial to obtain a clear picture of the basic reactions in such systems, and this is exactly what this project aims to achieve. Specifically, this project will initially focus on understanding the oxidation mechanism of magnesium with Quantum Espresso, an *ab initio* software package for electronic structure calculations of solids and is based on density functional theory. Adsorption studies are currently being carried out to study the surface behaviour as a function of hydroxide coverage on the magnesium close-packed surface. Similar calculations will be carried on a kink surface as well to establish a mechanism for dissolution of magnesium hydroxide to simulate a discharge cycle in the battery. The broader goal of this project is to develop a methodology to discern a trend in the behaviour of different metals in the same environment and summarize the results with a rationale on how better systems can be achieved.

(103b) Transient Interfacial Zinc Oxide Formation in Cycling Alkaline Batteries Detected Using Synchrotron X-Rays

Gallaway, J., City college of New York
Erdonmez, C., Brookhaven National Laboratory
Zhong, Z., Brookhaven National Laboratory
Croft, M.,
Sviridov, L., Energy Institute, City College of New York
Banerjee, S., City college of New York
Steingart, D., The City College of New York
Originally presented on: 11/4/2013 12:52:00 - 13:14:00

Energy dispersive X-ray diffraction (EDXRD) using white beam synchrotron radiation allows electrochemical cells with centimeter scale dimensions to be rapidly mapped crystallographically in both time and space. Using this technique, X-rays with a wide range of energies penetrate the cell, and X-ray diffraction intensity is measured at a single angle from a localized gauge volume within the cell. We report in situ EDXRD data collected on cycling LR20 Zn-MnO2 batteries collected using the superconducting wiggler beamline X17B1 at the National Synchrotron Light Source (NSLS) located at Brookhaven National Lab.

As data collection time for an EDXRD diffraction contour was short compared to the length of a battery cycle, material changes were resolved as a function of cycling stage as well as spatial location. This allowed us to resolve highly localized crystallographic changes within the battery. This is important, as batteries are designed to maximize interfacial area by employing thick, porous electrodes, which can display nonhomogenous reaction conditions. Concentration and reaction distributions within the electrodes can cause conditions that lead to battery failure during long term cycling.

The data revealed transient zinc oxide layers that formed and dissolved at the anode-separator interface during cycling, as a function of battery discharge rate. Massive zinc oxide formation was found in some cycled Zn-MnO2, and is believed to be a failure mechanism. Lattice strain was also detected in the cathode MnO2, and this lattice strain was found to be a function of radial distance from the anode.

It is generally recognized that electrochemical storage on a large scale is desirable for a number of societal reasons including electrical load leveling, power regulation, electrical demand peak shaving, the smart grid, and for firming intermittent renewables-based generation such as solar and wind power. Powerful in situ techniques such as EDXRD can aid in classifying long term battery failure mechanisms, which can have a multitude of causes some of which are localized and occur within thick, sealed devices.

(103c) Nonaqeous Electrolytes for Magnesium-Oxygen Batteries

Vardar, G., University of Michigan
Saraidaridis, J. D., University of Michigan
Sleightholme, A. E. S., University of Michigan
Siegel, D. J., University of Michigan
Monroe, C. W., University of Michigan
Originally presented on: 11/4/2013 13:14:00 - 13:36:00

Because of their high theoretical energy densities, metal-O_2 batteries have long been considered as an improvement upon current battery technologies. Mg-O_2 batteries have one of the highest predicted volumetric energy densities [1]. However, rechargeable Mg-O_2 batteries are not possible to realize with aqueous electrolytes due to non-conducting reaction products that form on the Mg anode. Non-aqueous Mg-conducting electrolytes have been studied in the context of Mg-ion and Mg-sulfur batteries [2]. Our goal is to develop a non-aqeuous electrolyte that is compatible with an O_2 cathode and stable for the MgO (2.95V vs. Mg/Mg^{++}) and MgO_2 (2.91V vs. Mg/Mg^{++}) formation reactions.

In this presentation, we will discuss our latest progress in electrolytes for rechargeable Mg-O_2 batteries. Figure 1 shows a cyclic voltammogram measured in magnesium borohydride ($Mg(BH4)_2$) in dimethoxyethane (DME). The oxidative stability of $Mg(BH4)_2$ in DME was found to be below 2V on a Pt working electrode. Interestingly, ionic liquids are known to have wide electrochemical windows. The oxidative stability of 1-Butyl-3-methylimidazolium bistrifluoromethylsulfonylimide (BmimTFSI) in DME was measured to be over 3V vs. Mg/Mg^{++} (Figure 2). However, addition of BmimTFSI to the $Mg(BH_4)$/DME solution reduced the reductive stability window considerably and Mg deposition/dissolution was not observed. Mg deposition and dissolution experiments are underway for non-aqueous systems with ionic liquids as the supporting electrolyte.

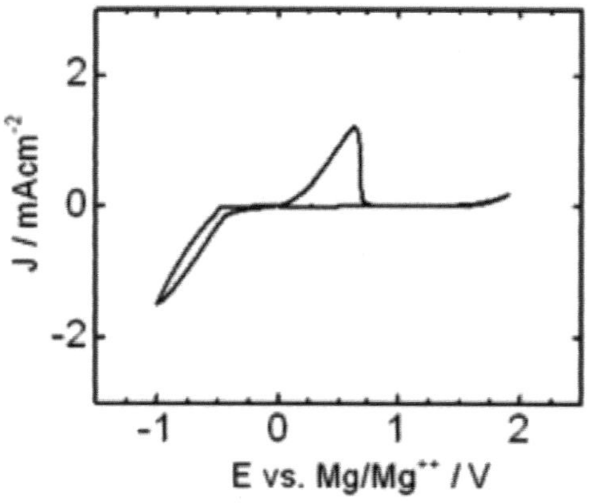

Figure 1. Cyclic voltammogram of Pt electrode in 0.05M $Mg(BH_4)_2$ in DME.

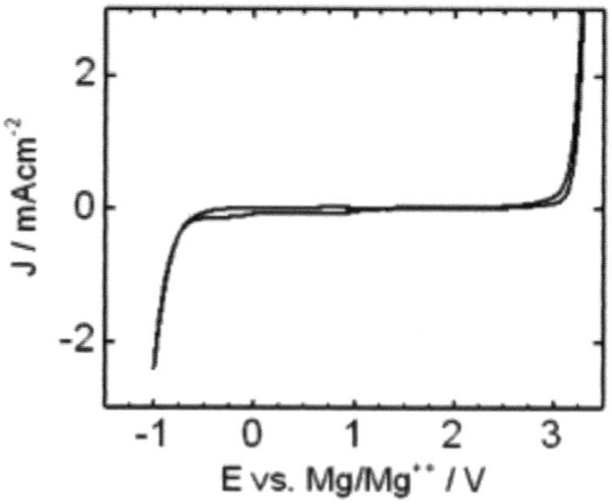

Figure 2. Cyclic voltammogram of Pt electrode in 0.1M BmimTFSI in DME.

[1] C.-X. Zu and H. Li, "Thermodynamic analysis on energy densities of batteries," Energy & Environmental Science, **4**, 2614, (2011).

[2] J. Muldoon et al., "Electrolyte roadblocks to a magnesium rechargeable battery," Energy & Environmental Science, **5**, 5941 (2012).

(103d) Incorporating Nano-Si Particles Into Cellulose-Based Anodic Composites for Lithium-Ion Batteries

Wokpetah, J., The Pennsylvania State University
Originally presented on: 11/4/2013 13:36:00 - 13:58:00

Incorporating Nano-Si Particles into Cellulose Fibers as a Way of Making Anodic Composites for Lithium-Ion Batteries

Joseph Wokpetah[1, 2], Simon Leijonmarck[2] and Goran Lindbergh[2]

[1]The Pennsylvania State University, Department of Chemical Engineering, 212 Fenske Laboratory, University Park, PA 16802,USA

[2] KTH-Royal Institute of Technology, Division of Applied Electrochemistry, Teknikringen 42, 100 44 Stockholm, Sweden

Lithium-ion batteries are a class of rechargeable batteries in which lithium ions migrate from the cathode to the anode during discharge and vice versa during charging of the battery. These batteries have become if not the most important, one of the important batteries over the years and have grown from R&D interest to sales of over 4 billion units in 2009 [3]. Most devices and electronics that require multiple recharging cycles use lithium ion batteries since they can be recharged many times without losing their capacity. In addition, due to their high electric potential, one can store lots of energy per mass of lithium.

One of the many areas of lithium-ion batteries that are still being researched is the anode material. Silicon is becoming more preferred compared to carbon as the negative electrode material due to many factors and the most renowned being its capacity: '3579 mAh/g for silicon versus 372 mAh/g for carbon' [4]. This means that using silicon instead of carbon, as the anodic material for lithium-ion batteries will make the battery better in terms of storage of energy density and cycling.

Silicon is, however, known to fall off the cupper current collector during discharging and charging of the battery, due to the increase in its size as it forms complexes with the lithium ions. This effect decreases the capacity and lifetime of the battery. Sanchez et al have shown that binding these silicon particles on the surface of a cellulose fiber increases the batteries' performance[5]. This project aims to investigate how lithium-ion batteries will perform when silicon particles are incorporated in the pores of a cellulose fiber used in making the anodic material for the battery.

The method used involves continuous mixing of cellulose fibers with silicon and carbon particles in ethanol. Part of the mixture is solvent exchanged from ethanol to acetone and finally to pentane, whilst the other part is not solvent exchanged. The purpose of solvent exchange is to keep the fibers relatively open after drying. We hypothesize that after analysis of our final results, the battery's capacity and lifetime will increase compared to the conventional method which uses just silicon.

We were successful in building a functional lithium ion battery with a silicon-cellulose anodic material but had challenges with the battery's cyclability.

---

[3] Edited by Thomas B. Reddy. Linden's Handbook of Batteries. Fourth edition. (2011)

[4] J.L. Gomez Camer, J. Morales and L. Sanchez. Nano-Si/Cellulose composites as anode materials for lithium-ion batteries. Electrochemical and solid-state Letters. (2008)

[5] J.L. Gomez Camer, J. Morales and L. Sanchez. Nano-Si/Cellulose composites as anode materials for lithium-ion batteries. Electrochemical and solid-state Letters. (2008)

(103e) Metal Complexes With Redox-Active Ligands As High Energy Density Nonaqueous Redox Flow Battery Electrolytes

Cappillino, P., Sandia National Laboratories
Pratt, H.,
Hudak, N.,
Tomson, N.,
Anderson, T.,
Anstey, M.,
Originally presented on: 11/4/2013 13:58:00 - 14:20:00

Metal Complexes with Redox-active Ligands as High Energy Density Nonaqueous Redox Flow Battery Electrolytes

by Patrick J. Cappillino, Harry D. Pratt III, Nicholas S. Hudak, Neil C. Tomson, Travis M. Anderson, Mitchell R. Anstey

Renewable energy sources that do not contribute to accumulation of greenhouse gases, such as wind and solar, must become an increasing portion of the global energy portfolio.Â While abundant, these sources are intermittent due to factors such as day-night cycles and weather.Â Furthermore, their generation capabilities are poorly matched to peak and off-peak use of electricity. Grid-scale strategies to mitigate these challenges, such as load-shifting, require energy storage components to supplement our current infrastructure in which electricity is instantaneously generated and distributed.

Redox flow batteries (RFB) have attractive features for grid-scale energy storage, including separation of energy and power capacities and modular scalability.Â Over the last several decades, a number of technologies based on vanadium and iron-chromium electrolytes were developed but have so far met limited commercial success. Further advancement requires fundamentally different redox chemistry with stable redox couples, large potential differences, and fast redox kinetics.

Herein we report new, non-aqueous RFB electrolytes that meet these criteria based on metal complexes containing ?redox non-innocent? ligands (M-NIL). The abundance of well-characterized, reversible charge transfer events exhibited by these compounds has the potential to greatly improve their energy density over state-of-the-art, transition-metal based RFB electrolytes.Â Details on their preparation, characterization and electrochemistry will be presented and their performance as RFB electrolytes will be discussed.

Sandia National Laboratories is a multi-program laboratory managed and operated by Sandia Corporation, a wholly owned subsidiary of Lockheed Martin Corporation, for the U.S. Department of Energy's National Nuclear Security Administration under contract DE-AC04-94AL85000.

(103f) Lithium Air Rechargable Battery Challenges to Sustained Cycling

Scott, K.,
Cheng, H.,
Sahapatsombut, U.,
Olubukun, O.,
Originally presented on: 11/4/2013 14:20:00 - 14:42:00

There is growing interest in with smaller and lighter weight batteries in combating global warming. Rechargeable Li batteries have been considered as serious contenders for hybrid electric vehicles and stationary power applications. Unfortunately, the energy density of current rechargeable lithium batteries is limited by the positive electrode, e.g. $LiCoO_2$ (130 $mAhg^{-1}$). A revolutionary advance from graphite-$LiCoO_2$ batteries to Li-air counterparts allows Li^+ and e^- in the cell to react with O_2 from the air. This enables an increase in storage capacities up to 10 times while reducing cost significantly. The air cathode is the new challenge for the battery. In this work, novel nanostructured air cathode materials were fabricated, characterised by using XRD, TEM and SEM and tested as cathodes in rechargeable Li-air batteries. Such 3D nanomaterials displayed notable successes in terms of capacity (up to 3000 $mAhg^{-1}$), power and materials sustainability. The superior performance of the nanomaterials was attributed to their structures which acted as transport pores and ensured rapid insertion and removal of lithium via the following reversible reaction:

$$2\ Li + O_2\ \ Li_2O_2$$

The relationship between composition, especially ratios of catalyst to carbon, structure, properties and performance shall be discussed. Measures to tackle technical barriers in battery design and engineering will also be addressed.

We have investigated the use of graphene as an alternative carbon conducting material in the Li-air battery using bifunctional MnO_2/Au catalysts. This has provided significant improvement in its stability (stable cycling > 100) as well as a great improvement in the lithium air cell capacity (10,000mAh/g).

Battery performance is still affected by loss in capacity after repeated cycling due to instability of certain solvents and cathode carbon supports. A micro-macro homogeneous mathematical model was developed for the cycling operation of a porous Li-air cathode using a concentrated binary electrolyte theory. The model includes mechanism for solvent degradation and production of Li carbonates. The published physical and chemical property data is carefully applied in this model to predict the time dependance of electrolyte concentration, non-uniform porosity and reaction rate and loss of capacity after cycling.

(103g) High Voltage Vanadium-Metal Hydride Hybrid Flow Battery As Electrochemical Energy

Li, C. Y.,
Weng, G., The University of Hong Kong
Chan, K. Y., The University of Hong Kong
Originally presented on: 11/4/2013 14:42:00 - 15:04:00

By coupling the advantages of the vanadium redox flow battery and the acid/alkaline hybrid battery concepts, a novel Vanadium-metal hydride (V-MH) hybrid rechargeable flow battery is developed. The V-MH flow battery consists of graphite felt positive electrode in flowing $VOSO_4$-H_2SO_4 electrolyte and metal hydride negative electrode in flowing KOH, the two flowing streams are separated by a bipolar membrane. Hybridizing the V^{4+}/V^{5+} coupled with metal hydride eliminates the problem of V^{2+} oxidation as in VRF battery. In the meantime, the higher voltage and higher capacity achieved is a result of the extra energy store/released in acid-alkaline neutralization corresponding to =-79.85 kJ/mol.

The charge and discharge electrode reactions in this acid/alkaline Vanadium-Metal Hydride flow system are as follows:

Graphite positive electrode:

$$VO_2^+(aq.) + 2H^+(aq.) + e^- \leftrightarrow VO^{2+}(aq.) + H_2O \quad (E^o=1.0\ V)$$

Metal Hydride negative electrode:

$$MH_x + OH^-(aq.) \leftrightarrow MH_{x-1} + H_2O + e^- \ (E^o= -0.8\ V)$$

These two electrodes are separated by a bipolar membrane, which provides ionic contacts and acts as barrier for proton and hydroxide ions transport in the pH differential cells.

Overall Discharge(®) and Charge(¬) reactions are

$$VO_2^+(aq.) + 2H^+(aq.) + MH_x + OH^-(aq.) \leftrightarrow VO^{2+}(aq.) + MH_{x-1} + 2H_2O \quad (E^o=1.8\ V)$$

In principle, this hybrid battery system of V^{4+}/V^{5+} and metal hydride can deliver an overall cell capacity of 110 mAh/g, cell voltage of 1.8 V, and a much higher energy density of 200 Wh/kg than that of the conventional all vanadium redox flow (VRF) battery (60.5 Wh/kg). The high voltage (1.8 V) reported is among the highest in the Vanadium-based hybrid flow systems, about 40% higher than conventional VRF battery (1.4 V) and nickel metal hydride battery (1.2 V). The electrochemical properties which includes the charge/discharge characteristics, cycle test and discharge polarization curves, of this new hybrid flow battery at different flow rates and electrolyte concentrations will be demonstrated and optimized. This V-MH flow battery has a huge potential to implement as effective electrochemical energy storage system from intermittent energy sources like wind and solar power.

The next stage of this V-MH battery would be to replace the metal-hydride electrode by hydrogen, resulting in a Vanadium-Hydrogen (V-H_2) flow battery. Since alkaline hydrogen electrode has an extremely high capacity and favorable reversible electrochemical kinetics, further increase the energy and power density of the system. In theory, this V-H_2 battery system can yield a cell voltage of 1.83 V, and reach a higher energy density of about 300 Wh/kg (which is five times that of VRF battery).

Acknowledgements

This research is supported by Innovation Technology Fund of Hong Kong (ITF ITS/290/12), Research Grants Council of Hong Kong (GRF HKU 700209P and GRF HKU 700210P), University of Hong Kong SRT, University of Hong Kong Small Project Funding, The University Development Fund on the Initiative of Clean Energy and Environment, and Zentric Inc.

References

[1] S.A. Cheng and K.Y. Chan*, "High Voltage Dual Electrolyte Electrochemical Power Source", *ECS Trans.* 25(35) (2010) 213-219.

[2] H. Li, G. Weng, C.Y.V. Li*, and K.Y. Chan*, "Three Electrolyte High Voltage Acid-alkaline Hybrid Rechargeable Battery", *Electrochim. Acta*, 56(25) (2011) 9420-9425.

[3] G. Weng, V.C.-Y. Li*, and K.Y. Chan*, "Lead Acid-NiMH Hybrid Battery System Using Gel Electrolyte", *ECS Trans.*41(13) (2012) 133-143.

[4] G. Weng, V.C.-Y.Li*, and K.Y. Chan*, "Study of the Electrochemical Behavior of High Voltage Vanadium-Metal Hydride Hybrid Semi-Flow Battery", *ECS Trans., to appear* (2013).

(134a) Identifying Defect Sites in Lithium-Ion Battery Materials: Local Disorder in LiVPO4F and Its Influence On Bulk Properties

Messinger, R. J., CNRS
Menetrier, M., CNRS
Carlier, D.,
Ateba Mba, J. M., CNRS
Croguennec, L.,
Masquelier, C.,
Massiot, D.,
Deschamps, M.,
Originally presented on: 11/4/2013 12:30:00 - 12:45:00

The transition metal fluorophosphate $LiVPO_4F$ is a promising electrode material for lithium-ion batteries that is capable of operating both as a high-voltage positive electrode (4.25 V, $V^{III/IV}$) and a negative electrode (1.8 V, $V^{II/III}$).[1,2] While its Tavorite-like structure has been solved by a combination of X-ray and neutron diffraction, solid-state nuclear magnetic resonance (NMR) measurements reveal a significant extent of local disorder that has not been observed at longer length scales with scattering techniques. In particular, solid-state single-pulse ^7Li NMR spectra acquired under conditions of ultra-fast (64 kHz) magic-angle-spinning (MAS) reveal that $LiVPO_4F$ exhibits several additional ^7Li NMR signals of unknown origin that can account for up to 20% of the total lithium content. Electrochemical measurements in Swagelok-type cells with lithium metal as the negative electrode establish that materials with different populations of unknown lithium sites exhibit different cycling behavior, including differing extents of irreversible capacity and polarization. 2D $^7Li\{^7Li\}$ dipolar correlation NMR measurements establish the sub-nanometer proximities of the unknown lithium sites with respect to the $LiVPO_4F$ site, proving unambiguously that the unknown sites are local defects within the $LiVPO_4F$ structure and are not associated with separate impurity phases. In combination with magnetic measurements, variable-temperature NMR measurements reveal that the paramagnetic ^7Li NMR shifts of the different lithium sites vary linearly with the bulk magnetic susceptibility, further confirming that the sites are defects within the $LiVPO_4F$ structure. The results provide a general strategy for identifying lithium defect sites in lithium-ion battery materials and highlight the importance of understanding local disorder and its relationship to macroscopic material and device properties.

[1]J.-M. Ateba Mba, C. Masquelier, E. Suard, L. Croguennec, *Chem. Mater.* **2012**, *24*, 1223 – 1234.

[2]J.-M. Ateba Mba, L. Croguennec, N.I. Basir, J. Barker, C. Masquelier, *J. Electrochem. Soc.* **2012**, *159*, A1171-A1175.

(134b) Electrochemical Properties and Characterization of Li1.2Mn0.54Co0.13Ni0.13O2 Composite Cathode Powders Prepared by Ultrasonic Spray Pyrolysis

Langrock, A., University of Maryland
Ehrman, S. H., University of Maryland
Wang, C., University of Maryland
Originally presented on: 11/4/2013 12:45:00 - 13:00:00

Lithium-rich layered oxide $Li_{1.2}Mn_{0.54}Co_{0.13}Ni_{0.13}O_2$ (LMO) composite cathode powders are prepared by ultrasonic spray pyrolysis. The crystal structure and morphology of the prepared samples are characterized with X-ray diffraction and scanning electron microscope. The spherical shape and porous nanostructure of the as prepared powders are maintained after post-treatment at temperatures between 600 and 800°C. Grain growth in the powders increases with increasing post-treatment temperatures. The BET surface area of the composite cathode powders post-treated at 800°C is 12 m^2 g^{-1}. Discharge capacity of the cathode powders post-treated at 800°C decreases from 278 mAh g^{-1} to 208 mAh g^{-1} by the 50th cycle, the capacity retention being 75%. The post-treatment temperature strongly affected the electrochemical properties of the composite cathode powders. Composite cathode powders post-treated at high temperatures with high amounts of the Li_2MnO_3 phase had good cycle properties as well as large initial charge/discharge capacities.

(134c) Atomic Layer Deposition (ALD) On the Nanostructured Li-Mn-Rich Composite Li1.2Ni0.13Mn0.54Co0.13O2 Cathode Powder

Zhang, X., Institute of Nuclear and New Energy Technology, Tsinghua University
Belharouak, I.,
Originally presented on: 11/4/2013 13:00:00 - 13:15:00

Atomic Layer Deposition (ALD) on the Nanostructured Li-Mn-rich Composite $Li_{1.2}Ni_{0.13}Mn_{0.54}Co_{0.13}O_2$Cathode Powder

Xiaofeng Zhang,[a] Ilias Belharouak,[a,*]

[a] Chemical Sciences and Engineering Division, Argonne National Laboratory, 9700 S. Cass Ave, Argonne, IL 60439, USA, Email: belharouak@anl.gov(Corresponding author)

Li-Mn-rich (LMR) composite cathode materials $Li_{1+x}M_{1-x}O_2$ (M = Ni, Mn and Co) are of particular interest for plug-in hybrid electric vehicles (PHEVs), in which a high-energy battery is always desired for a higher mileage per charge.[1, 2] In the composite structure, the layered, the spinel and rocksalt structure are compatible at nano-scales, that many composites with different electrochemical properties can be synthesized by simply tuning the Li content in the materials. However, these materials would suffer from voltage decay and capacity fading with cycles, which significantly deteriorate their energy densities and life time for practical applications. In this type of materials, Li content is essential to the arrangement of the transition metal atoms in the composites and also to the structural stability against undesired reactions in a real electrochemical cell. It has been demonstrated that the occurrence of layered-spinel transformation induces the voltage decay. It is because the Li intercalation/deintercalation reactions may preferentially occurs at the tetrahedral site other than octahedral site, which facilitates the spinel formation.[1, 2] Further study by HR-TEM indicated that the composite particle surface could lose the layered feature after a few cycles, and the first-principle calculation suggested a defect-spinel structure forming on the surface due to the migration of transition metal ions into the Li layers.[3] Therefore, the surface chemistry and structure stability appear to be essential to limit the undesired reactions and phase transformation in the composite materials. Some achievements have been made *via* developing new synthesis processes, new chemistry or surface modifications (e.g., coating), which possibly altered the short-range and long-range ordering of transition metal atoms (Ni and Mn) near the surface, which lead to an improvement in the battery performance.[4-8]

Atomic layer deposition (ALD) is a well-established method to grow ultrathin films on flat surfaces and powders in atomic thickness using sequential, self-limiting surface reactions.[9-14] It has been demonstrated that the surface film can react as a surface barrier between the cathode active material and the electrolyte, which would reduce the loss of active materials in the electrode. [9-14] Consequently, less capacity loss is achieved for the coated cells as compared with the pristine materials. The well-established ALD process normally is using trimethylaluminium (TMA) and H_2O as precursors forming an Al_2O_3 surface film. Experimental results showed that the coated $LiCoO_2$ powders exhibited a capacity retention of 89% after 120 charge–discharge cycles in the 3.3–4.5 V vs. Li/Li+ range. [12, 15-17] The ALD surface coating can also be applied to battery electrodes at a low temperature <150 ºC, which would not degrade the electrodes (e.g., melting of polymer binders) in this temperature range. It is reported that Al_2O_3-coated $LiCoO_2$ electrodes display superior electrochemical performance in comparison to the uncoated $LiCoO_2$ electrodes. However, Liu et al observed that the ALD surface coated Al_2O_3 could be lithiated forming passivation $LiAlO_2$ glass on the materials surface using in-situ TEM [18], which provided new insight into the electrochemical characteristic of the Al_2O_3surface film against Li.

In this paper, we are aiming for studying the importance of a conformal surface protection film using atomic layer deposition (ALD) technique on the high-capacity $Li_{1.2}Ni_{0.13}Mn_{0.54}Co_{0.13}O_2$ cathode materials in Li-ion batteries, which can be potentially significantly improved in their electrochemical performance. More precisely, we evaluated the effect of coating species, coating thickness and morphology on the rate performance, cycling life and high temperature performance of the high-capacity $Li_{1.2}Ni_{0.13}Mn_{0.54}Co_{0.13}O_2$ cathode materials in half-cell and full-cell configurations.[19] We found that an conformal Al_2O_3 thin film could be deposited on the powders using 10 ALD cycles of TMA and H_2O at 150 °C. TiO_2 coating was prepared using 20 ALD cycles of titanium isopropoxide (TTIP) and H_2O at 150 °C, which displayed a particulate morphology. Nevertheless, the surface coated TiO_2 film was electrochemically active, which could be lithiated forming Li_xTiO_2 as detected by X-ray photoelectron spectroscopy (XPS). The electrochemical reactivity of TiO_2 will cause degradation of the surface film, leading to capacity fading of the cell. Nevertheless, it was also found that the Al_2O_3 surface coating was not electrochemically active after repetitive cycles, which was contrary to the early report by Lui et al that Al_2O_3 can be lithiated into $LiAlO_2$. [18] To further explorate the phenomena, a systematic study on the ALD coating of Al_2O_3 and $LiAlO_2$was performed.

Refereances:

[1] C.S. Johnson, N. Li, J.T. Vaughey, S.A. Hackney, M.M. Thackeray, Electrochem. Commun., 7 (2005) 528-536.

[2] C.S. Johnson, N.C. Li, C. Lefief, J.T. Vaughey, M.M. Thackeray, Chem. Mater., 20 (2008) 6095-6106.

[3] B. Xu, C.R. Fell, M.F. Chi, Y.S. Meng, Energy & Environmental Science, 4 (2011) 2223-2233.

[4] J. Liu, B. Reeja-Jayan, A. Manthiram, Journal of Physical Chemistry C, 114 (2010) 9528-9533.

[5] S.H. Kang, M.M. Thackeray, J. Electrochem. Soc., 155 (2008) A269-A275.

[6] J.R. Croy, M. Balasubramaniam, D. Kim, S.-H. Kang, M.M. Thackeray, Chem. Mater., 23 (2012) 5415-5424.

[7] J.R. Croy, S.H. Kang, M. Balasubramanian, M.M. Thackeray, Electrochem. Commun., 13 (2011) 1063-1066.

[8] D. Kim, S.H. Kang, M. Balasubramanian, C.S. Johnson, Electrochem. Commun., 12 (2010) 1618-1621.

[9] M.Q. Snyder, S. Trebukhova, B. Ravdel, M.C. Wheeler, J. DiCarlo, C.P. Tripp, W.J. DeSisto, Mater Res Soc Symp P, 972 (2006) 331-338.

[10] M.Q. Snyder, S.A. Trebukhova, B. Ravdel, M.C. Wheeler, J. DiCarlo, C.P. Tripp, W.J. DeSisto, J. Power Sources, 165 (2007) 379-385.

[11] S.M. George, Chem. Rev., 110 (2010) 111-131.

[12] Y.S. Jung, A.S. Cavanagh, A.C. Dillon, M.D. Groner, S.M. George, S.H. Lee, J. Electrochem. Soc., 157 (2010) A75-A81.

[13] L.A. Riley, A.S. Cavanagh, S.M. George, S.H. Lee, A.C. Dillon, Electrochem. Solid-State Lett., 14 (2011) A29-A31.

[14] L.A. Riley, S. Van Ana, A.S. Cavanagh, Y.F. Yan, S.M. George, P. Liu, A.C. Dillon, S.H. Lee, J. Power Sources, 196 (2011) 3317-3324.

[15] Y.S. Jung, A.S. Cavanagh, L.A. Riley, S.H. Kang, A.C. Dillon, M.D. Groner, S.M. George, S.H. Lee, Advanced Materials, 22 (2010) 2172-2176.

[16] I.D. Scott, Y.S. Jung, A.S. Cavanagh, Y.F. An, A.C. Dillon, S.M. George, S.H. Lee, Nano Lett., 11 (2011) 414-418.

[17] Y.S. Jung, P. Lu, A.S. Cavanagh, C. Ban, G.H. Kim, S.H. Lee, S.M. George, S.J. Harris, A.C. Dillon, Advanced Energy Materials, 3 (2012) 213-219.

[18] Y. Liu, N.S. Hudak, D.L. Huber, S.J. Limmer, J.P. Sullivan, J.Y. Huang, Nano Lett., 11 (2011) 4188–4194.

[19] X. Zhang, I. Belharouak, L. Li, Y. Lei, J.W. Elam, A. Nie, X. Chen, R. Shahbazian-Yassar, R.L. Axelbaum, Advanced Energy Materials, (2013). 10.1002/aenm.201300269

(134d) Metal Fluoride/Graphene Composites for High-Performance Li-Ion Cathodes

Hayner, C. M., Northwestern University
Zhao, X., Northwestern University
Yu, Y. Y., Northwestern University
Kung, M. C., Northwestern University
Kung, H. H., Northwestern University
Originally presented on: 11/4/2013 13:15:00 - 13:30:00

Rechargeable batteries with improved energy and power density have many potential, high impact applications including advanced portable electronics and electric vehicles. However, most 'next-generation' materials capable of providing improved performance suffer from rapid capacity degradation or severe loss of capacity when rapidly discharged. Graphene, a unique two-dimensional material composed of atomically thin carbon sheets, has shown potential to address unsatisfactory rate capability, limited cycling performance and abrupt failure of these high capacity material. Therefore, we have been exploring graphene-based architectures, allowing us to exploit graphene's unique combination of high surface area, high in-plane electrical conductivity, excellent tensile modulus and mechanical durability.

Metal fluorides, including iron trifluoride (FeF_3), have the potential of both high energy delivery and high cyclability. We have recently developed a stable nanostructured iron fluoride (FeF_3)/graphene composite cathode. A facile route that combines co-assembly and photothermal reduction was developed to synthesize free-standing, flexible FeF_3/graphene papers. The papers contained a uniform dispersion of FeF_3 nanoparticles (< 40 nm) and open ion diffusion channels in the porous, conducting network of graphene sheets that resulted in a flexible paper cathode with high charge storage capacity, rate, and cycling performance, without the need for other carbon additives or binder. The composites demonstrate promising applications as cathodes in high-energy density Li-ion batteries. In a free-standing form, they showed a high storage capacity of >400 mAh/g and improved cycling performance compared to bare FeF_3 particles. In addition, we have investigated a novel mixed metal fluoride cathode system, containing tunable manganese and iron fluoride, for which characterization and electrochemical performance will be presented.

(134e) Nanostructured Li2S-Graphene Composites As Cathode for High Energy Density Li-S Batteries

Han, K., Northwestern University
Shen, J., Northwestern University
Hayner, C. M., Northwestern University
Yu, Y. Y., Northwestern University
Kung, M. C., Northwestern University
Kung, H. H., Northwestern University
Originally presented on: 11/4/2013 13:30:00 - 13:45:00

Lithium-ion batteries, although a promising technology for energy storage, is reaching their energy density limitation, making it difficult to fulfill the requirements of emerging applications such as electric vehicles and smart grid. Lithium-sulfur batteries are one of the most attractive systems for next generation high energy storage beyond Li-ion batteries. Conventional Li-S batteries mainly use elemental sulfur as the starting cathode material due to its high theoretical capacity of ~1675 mAh/g and high energy density of ~2500 Wh/kg based on the redox reaction $S_8 + 16Li \rightleftharpoons 8Li_2S$. Sulfur is also low cost, abundant, and of low toxicity. However, significant hurdles such as rapid capacity fade and poor rate capability have to be overcome before practical application can be realized. Two important underlying reasons for the poor performance of this battery are the insulating nature of sulfur and the dissolution of the intermediate product polysulfide into the electrolyte resulting in redox shuttling. Although extensive research efforts have been expanded to solve these issues, most studies are limited to the simple cell configuration of sulfur as cathode and metallic lithium as anode. For safety and ease of operation, introducing Li into the cell in the form of Li_2S is more desirable than Li metal. Moreover, Li_2S, with a theoretical capacity of ~1166 mAh/g, could be paired with a lithium-metal-free anode such as Si or Sn.

Here, we report the use of a nanostructured Li_2S-graphene composite as cathode material. Li_2S nanoparticles (20~50nm) are imbedded in pocket-like structures in a three-dimensional graphene/reduced graphene oxide via a facile, wet chemical method followed by thermal treatment. Structural characterization by XRD and SEM provide clear evidence for the unique structure. The constructed 3-D graphene network could enable efficient use of Li_2S active material by providing excellent electrical contacts, offer long-term stability against loss of polysulfide and accommodate volume changes during charge and discharge processes. The as-prepared Li_2S-graphene nanocomposites exhibit high initial reversible capacity and good cycling performance. The effect of lithium polysulfide as additive of the electrolyte is also investigated, especially with respect to its contribution to the overall charge storage capacity. The structural and electrochemical characterization results of the nanocomposites will be presented.

(134f) Carbon/Sulfur Microspheres With Multi-Modal Pore Structures for Lithium-Sulfur Battery Cathodes

Zhao, C., University of Wisconsin-Milwaukee
Liu, L., University of Wisconsin-Milwaukee
Zhao, H., University of Wisconsin-Milwaukee
Li, Y., University of Wisconsin-Milwaukee
Originally presented on: 11/4/2013 13:45:00 - 14:00:00

Sulfur has received increasing attention as a cathode material for lithium-sulfur batteries due to its high theoretical specific capacity of 1675 mAh g-1, which is five times higher than that of conventional lithium-ion batteries. However, poor electronic conductivity of sulfur, dissolution of the lithium polysulfide intermediates and large volume expansion of sulfur during cycling limit the commercialization of Li-S batteries. In this work, we have demonstrated a cathode design with a novel structure made of multi-modal porous carbon microspheres. The tunable porous structure of sulfur-infiltrated carbon nanocomposites allows easy access to Li ion transport, inhibits loss of active sulfur material, and provides void space for sulfur expansion. An initial capacity of 1289 mAh g-1 at 1C and a Coulombic efficiency of 90% over 200 cycles are achieved.

(134g) Nanofiber-Based Cathode Electrocatalysts Supported On Carbon Substrates for Lithium-Air Rechargeable Batteries

Joo, Y. L., Cornell University
Yin, J., Cornell University
Kim, J.,
Originally presented on: 11/4/2013 14:00:00 - 14:15:00

Nanofiber-Based Cathode Electrocatalysts Supported on Carbon Substrates for Lithium-Air Rechargeable Batteries

*Jun Yin[a], Jangwoo Kim[a], and Yong Lak Joo[a]**

[a] School of Chemical and Biomolecular Engineering, Cornell University, Ithaca, New York 14853

In terms of usable energy density rivaling that of gasoline, rechargeable lithium-air batteries have attracted increasing interests presently. However, the practical capacity was highly affected by the effective catalyst materials in the air cathode in Li-air batteries. The demand for effective catalyst materials in the air cathode requires the design and synthesis multifunction catalysts for power density improvement. In this presentation, several types of metal oxide nanofibers (e.g., ZnO, Co_3O_4 etc) were synthesized by water-based electrospinning of highly loaded precursor solutions, followed by thermal treatment and their use as the cathode catalysts in single-cell measurements was described. In the meantime, the correlation between the various carbon-based substrates and the metal oxide fiber catalysts and how they synergistically influence the electrocatalytic activity will also be discussed. The metal oxide nanofibers are characterized using an array of techniques such as SEM, XRD, XPS, BET and Raman to determine their morphology and structural properties. Charging – discharging evaluation, cyclic voltammetric measurements in organic electrolytes were carried out to assess the mechanistic details as well.

(134h) Low-Temperature Dehydrogenation From Aminoborane Complexes Under Carbon Dioxide Atmospheres

Lee, J. W., The City College of New York
Xiong, R., CCNY
Zhang, J., The City College of New York
Originally presented on: 11/4/2013 14:15:00 - 14:30:00

Ammonia borane (AB) is a chemical hydride and one of attractive solid-state media for on-board hydrogen storage due to high hydrogen contents (19.6 wt. %) and low-temperature dehydrogenation compared to other metal hydrides. The hydrogen release from AB can be accelerated at the working temperature (~ 85 °C) of polymer electrolyte membrane fuel cell. However, this talk shows that CO_2-treatment can accelerate hydrogen release of AB polymer complex, polyaminoborane (PAB). Polyaminoborane (PAB) is the spent fuel of ammonia borane (AB, NH_3BH_3) thermolysis with around 1 mole hydrogen release per mol AB. Dehydrogenation of PAB by thermolysis generally commences at temperatures above 120 °C. However, we find gaseous CO_2 environments dramatically enhance the thermolytic dehydrogenation of polyaminoborane (PAB) at 85 - 90 °C. By exposing PAB to 2.07 MPa CO_2, 5.48 wt % hydrogen based on the mass of ammonia borane (AB) desorbs in less than 1 h, while 4 h is required to generate 5.04 wt % hydrogen by maintaining PAB at 120 °C without using CO_2. The total hydrogen desorbed from the initial AB is 11.38 wt % under the CO_2 atmosphere at 85 °C, including the hydrogen release during the transition from AB to PAB. The main contributing factors to the facile hydrogen release are the promoting effect of newly formed formic acid and the exothermic effect of the reaction between PAB and CO_2. We will explain the possible promoting mechanism by various spectroscopic and elemental analyses.

(134i) Intermittent Electrical Energy Storage By Liquid Organic Hydrogen Carriers

Arlt, W., University of Erlangen-Nuremberg
Muller, K., University of Erlangen-Nuremberg
Teichmann, D., University of Erlangen-Nuremberg
Originally presented on: 11/4/2013 14:30:00 - 14:45:00

The intermittent delivery of energy from sources like photovoltaics or wind demands a large and long-term storage of energy. Liquid Organic Hydrogen Carriers like "Carbazol" address this topic using the infrastructure of liquids like domestic fuel oil. In Germany, the situation is far more complicated, because the country decided to phase out nuclear energy in the year 2022. The lecture highlights the situation in the German state of Bavaria and the planning.

The lecture reports the progress of Liquid Organic Hydrogen Carriers (LOHC). A demo-project is under construction at a condo in the city of Erlangen/Bavaria. Important data are safety, efficiency of electrical energy storage and robust construction. The purpose of this demo is to investigate

1. The ratio between storage and heat demand of the building
2. The number of storage and delivery actions
3. The robustness of the process in a public area

The time frame is 2 years from beginning of 2014 on.

(134j) Synthesis of Nanoporous Palladium Powder With Controlled Pore and Particle Size for Hydrogen Storage Applications

Cappillino, P., Sandia National Laboratories
Jones, C.,
Hattar, K., Sandia National Laboratories
Clark, B.,
Hekmaty, M.,
Jacobs, B.,
Robinson, D. B., Sandia National Laboratories
Originally presented on: 11/4/2013 14:45:00 - 15:00:00

Synthesis of Nanoporous Palladium Powder with Controlled Pore and Particle Size for Hydrogen Storage Applications

by Patrick J. Cappillino, Christopher G. Jones, Khalid M. Hattar, Blythe G. Clark, Michelle A. Hekmaty, Benjamin W. Jacobs, and David B. Robinson

Materials capable of rapidly storing and delivering hydrogen are currently in demand for numerous energy applications.Â The high surface area exhibited by nanoporous palladium has the potential to greatly improve its kinetics of hydrogen charge and discharge *versus* nonporous metal while retaining favorable operating temperatures, pressures, and volumetric capacity.Â In addition, the void space imparted by porosity should accommodate the volumetric expansion of palladium upon hydriding, mitigating plastic deformation and improving cycle life.

By chemically reducing palladium salts in the presence of various surfactants, we are able to synthesize palladium having size-tunable mesopores from a few to tens of nm, in a scalable fashion.Â Our results suggest that Pd nanoparticles form and sinter around micelles present in the aqueous media.Â Changing the chemical composition of the surfactant affects the size of these micelles and ultimately determines the pore dimensions.Â We demonstrate that pore size affects not only surface area, but also pore thermal stability under vacuum, and in the presence of hydrogen gas.

In addition to pore geometry, particle size and shape are important factors in determining gas flow characteristics. Using nonporous copper particles as sacrificial reductants for palladium salts, in the presence of surfactants, we are able to synthesize micron-sized particles of uniform size and shape.Â This is in contrast to continuous aggregates and films that are obtained using soluble chemical reductants or planar electrodes.

We will present details of the synthesis of nanoporous palladium, demonstrating control of both pore and particle size.Â Bulk measurements of hydrogen storage properties and pore characteristics, as well as microscopic measurements, including *in situ* heated-stage TEM under vacuum and in the presence of hydrogen, will be included, as well as details on the kinetics of hydriding/dehydriding.

Sandia National Laboratories is a multi-program laboratory managed and operated by Sandia Corporation, a wholly owned subsidiary of Lockheed Martin Corporation, for the U.S. Department of Energy's National Nuclear Security Administration under contract DE-AC04-94AL85000.

(157a) Fuel Processing of Logistic Fuels for SOFC-Based APU Systems

Zhu, T., United Technologies Research Center
Tew, D.,
Junker, S. T.,
Hawkes, J. R.,
Fokema, M. D.,
Simmers, R., ASPEN PRODUCTS GROUP, INC
Originally presented on: 11/4/2013 15:15:00 - 15:37:00

This work describes the processing of logistic fuels (e.g.. JP-8/kerosene, DF-2/Diesel), for 1-10 kWe solid oxide fuel cell (SOFC) based auxiliary power units (APU). The fuel processing system consists of a desulfurizer, a catalytic reformer and a burner.

The sulfur content in JP-8, for instance, can be as high as 3000 ppm by weight (ppmw). In SOFC systems, effective desulfurization is essential for the performance of downstream catalytic reformer and stack(s). The system described herein uses a dual bed regenerable desulfurizer, developed by Aspen Products Group, Inc., to reduce the sulfur level in the inflowing liquid fuel from levels as high as 3000 ppmw to a level of less than 15 ppmw. The absorbents utilized in the desulfurization process are fully regenerable in situ during system operation. The desulfurizer consists of two absorbent beds, with one working in desulfurization mode while the other one working in regeneration mode. The system operates fully automated as it cycles between regeneration and desulfurization modes in order to maintain a constant flow of desulfurized fuel to the reformer. In this presentation, the effect of operating conditions on sulfur removal efficiency and fuel loss of the desulfurizer will be discussed. The general control strategy of the desulfurizer, especially during bed switching, and some testing results will be presented.

The desulfurized fuel is fed to an autothermal reformer (ATR) where it reacts with air and steam to generate CO and H_2 for the SOFC stack. A portion of the anode exhaust stream is recycled to the reformer to provide the steam required for the reformer. The reformer performance was evaluated using desulfurized fuel, air and a simulated anode recycle stream. The reforming efficiency decreases with increasing oxygen to carbon (O_2/C) ratio. The effect of steam to carbon (S/C) ratio is more complicated as the reformer efficiency initially increases with increasing S/C ratio and then decreases after S/C reaches 2.0. The optimum operating condition of the reformer is also determined by the formation of the olefins. The olefin products in reformate are detrimental to stack performance due to coke formation, whose formation under different operating conditions will be discussed. The reforming efficiency and pressure drop tradeoffs with different pore size ceramic foam catalysts will also be presented.

This work was supported by the U.S. Army's Tank Automotive Research Development and Engineering Center (TARDEC) with Program Managers Jeff Ratowski and Kevin Centeck. This support is gratefully acknowledged.

(157b) Designing Artificial Photosynthesis Cells Under Various pH Conditions

Berger, A. D., Pennsylvania State University
Modestino, M. A.,
Walczak, K.,
Segalman, R. A., University of California at Berkeley
Newman, J., UC Berkeley
Originally presented on: 11/4/2013 15:37:00 - 15:59:00

Using a photoelectrochemical cell (PEC) to produce solar fuels is sometimes called artificial photosynthesis, and it has attracted significant research interest [1-3]. The concept employs a photoactive component that absorbs light to produce a difference in potential. By interacting with an electrolyte, this potential difference can drive water electrolysis, possibly augmented with other fuel-producing reactions.

The design of such a system necessitates close interaction between experiment and modeling. The coupled processes of light absorption, electronic transport, catalysis, and solution transport must all be taken into account simultaneously. To this end, an integrated 1-D model is developed to predict device performance accurately. For geometries of interest, we specify target materials properties, such as the exchange current densities of catalysts and the conductivity of the membrane. For instance, in a microstructured device such as a forest of microwires embedded in a membrane, the membrane conductivity can be as low as 10^{-4}S/cm with minimal impact on device performance. The targets indicate that new membranes designed to replace Nafion may still improve device performance despite having lower ionic conductivity, as long as the new membranes also have lower gas permeability.

Because corrosion remains a significant limitation for photoelectrochemical cells (PECs), there is some desire to operate closer to neutral pH conditions. However, running under milder conditions presents a host of new problems that must be addressed. Two problems that have been shown experimentally are ohmic drop and losses due to pH gradients [4]. This model offers insight into the origin of these problems and how to avoid them. The choice of membrane and electrolyte composition affects the depletion of the electrolyte and, by extension, the limiting current behavior. Additionally, the relative magnitude of losses due to ohmic drop, the dependence of open circuit potential on pH, and the diffusion for different choices of electrolyte is calculated. Results show that a supporting electrolyte fails to eliminate the pH-dependent loss.

To overcome this, one strategy is to recirculate the electrolyte between the anode and cathode chambers. Experimental results are given for a membrane electrolysis cell operating with Nafion and aqueous potassium borate at 10 mA/cm^2 and a modest recirculation rate on the order of 60 mL/h. The model is in agreement with several experimental findings, including the decrease in pH difference between anode and cathode as the recirculation rate is increased, the current efficiency, and the composition of the exit stream. A maximum flow rate for each operating current density is identified in order to avoid flammable gas mixtures, and the optimum flow rate for overall hydrogen production efficiency is calculated.

References:

1. A. Listorti, J. Durrant, and J. Barber, Nat. Mater., 8, 929–930 (2009).

2. M. G. Walter, E. L. Warren, J. R. McKone, S. W. Boettcher, Q. Mi, E. A. Santori, N. S. Lewis, Chem. Rev., 110, 6446–6473 (2010).

3. S. Bensaid, G. Centi, E. Garrone, S. Perathoner, and G. Saracco, ChemSusChem, 5, 500–521 (2012).

4. E. A. Hernández-Pagán, N. M. Vargas-Barbosa, T. Wang, Y. Zhao, E. S. Smotkin, and T. E. Mallouk, Energ. Environ. Sci., 5, 7582 (2012).

(157c) Atomic Layer Deposition of Yttria-Stabilized Zirconium Using Cp-Type Precursors and Water Vapor for Fuel Cell Applications

Rossero, J. I., University of Illinois at Chicago
Takoudis, C. G., University of Illinois at Chicago
Klie, R. F., University of Illinois at Chicago
Gulec, A.,
Originally presented on: 11/4/2013 15:59:00 - 16:21:00

Thin films of yttria-stabilized zirconium oxide (YSZ) were successfully deposited via atomic layer deposition (ALD) using metal CP-type precursors Tris(isopropyl-cyclopentadienyl)yttrium and tris(dimethylamino)cyclopentadienylzirconium with water as the oxygen donor for use in solid oxide fuel cells (SOFC's). In order to characterize the ALD process, the growth rate of individual oxides was measured as function of precursor dosage, water vapor dosage, number of cycles and reactor temperature in a custom-designed and built hot wall ALD reactor. It was found that the ZrO_2 process has a growth rate of ~0.7 Å/cycle while for Y_2O_3 the growth rate is ~1.4 Å/cycle. An overlap of the individual ALD windows was found from 250 to 280°C; this overlap is ideal for the ALD of the composite YSZ films. The normalized cycle ratio defined as the cycles of yttrium precursor/total cycles was changed to investigate the tunability of the process. X-ray photoelectron spectroscopy (XPS) analysis was used to evaluate the composition of as-deposited YSZ. It was found that the normalized cycle ratio and the yttrium atomic percentage (Y atoms / metal atoms) have a linear relationship implying excellent tunability. Transmission electron microscopy (TEM) along with electron energy loss spectroscopy (EELS) was performed to characterize the crystalline structure and to determine the stoichiometry of the deposited films. TEM results show that the atomic percentage of yttrium in the YSZ films strongly influence the film crystallinity. For instance, films with 8% atom of Yttrium are polycrystalline with small amorphous grains while films with an atomic % of yttrium higher than 10% are polycrystalline with no amorphous regions.

(157d) Selective Ion Transport in Microporous Inorganic Membranes and Potential Applications As Battery Electrolyte Membranes

Michos, I., University of Cincinnati
Xu, Z., University of Cincinnati
Dong, J., University of Cincinnati
Gu, X., State Key Laboratory of Materials-Oriented Chemical Engineering, Nanjing University of Technology
Originally presented on: 11/4/2013 16:21:00 - 16:43:00

Redox-Flow Batteries (RFBs) have gained immense attention because of their unique advantages such as separate power and storage, low cost, and small environmental impact. Commercial demonstrations of large scale RFB units as energy storage devices for renewable power systems and large-scale grids are now being carried out in many countries around the world. However, fundamental researches are still needed to improve the cell reliability, energy efficiency and cost effectiveness. The ion exchange membrane (IEM), a central piece of the RFB, is currently a main subject of research because all the IEMs are made of ionic polymers which have the issues of relatively low ion selectivity and long-term material instability. This presentation reports our investigation of ion selective diffusion through nanoporous ceramic membranes and its potential use as IEMs in RFBs. We will discuss the ion permeability and selectivity of the membranes and their performance as IEM in RFB in terms of power density, energy efficiency, and charge-discharge behavior.

(157e) Simulations and Analysis of Fuel Cell Catalyst Degradation for the Development of Standardized Test Protocols

Raman, S., Texas Tech University
Maddala, J., Texas Tech University
Bullecks, B., Clarkson University
Rengasamy, R., Texas Tech University
Originally presented on: 11/4/2013 16:43:00 - 17:05:00

Fuel cells show promise as green energy and zero emission power source for several applications including transportation and stationary power production. Yet, the cost associated with energy production using fuel cells is higher than that of the conventional fossil fuels based technologies, which limits their commercial success. Platinum is the most widely used catalyst in fuel cell systems and it suffers from rapid degradation and depletion from the membrane electrode assembly. The degradation of catalysts can occur by several mechanisms namely dissolution, agglomeration, particle breakdown etc. Developing fuel cell catalysts with improved resistance to these degradation methods is a major challenge. However, tools to characterize the long term stability and durability of these new catalysts are not fully developed. In order to characterize the long term durability of catalyst materials and particles, there is a need to develop reliable and fast characterization techniques, using which new catalyst can be evaluated expeditiously. This will accelerate the development of new catalysts with much better long term reliability. Theories in literature consider several mechanisms of catalyst degradation and quantify them at certain operating voltages. Traditionally degradation is characterized through cyclic voltammetry (CV) over several thousand sweep cycles of voltage.

In this work, we simulate the degradation of platinum catalysts in the presence of various voltage cycle patterns utilizing the kinetics of the platinum degradation reaction. The parameters of the CV model are obtained by comparing the model with experiments. The degradation mechanisms that we consider are dissolution and oxide formation in platinum. This model is also extended to include the effect of agglomeration to study the particle size distribution using Oswald ripening dynamics. The use of the proposed approach in developing test protocols for rapid screening of new catalysts will be highlighted.

(157f) Mathematical Model for Electrochemical Insertion of Lithium in Nanostructured Silicon Electrodes

De, S., Washington University in St. Louis
Sikha, G.,
Gordon, J.,
Originally presented on: 11/4/2013 17:05:00 - 17:27:00

Abstract:

Silicon, germanium etc. are presently being pursued as potential anode materials for lithium-ion batteries owing to their high gravimetric (mAh/g) and volumetric capacities (mAh/L) compared to existing state of art graphite, for high energy and high power applications of the future [1]. One of the critical challenges is to overcome particle fracture resulting from internal stresses, developed during lithium intercalation. Recent experimental studies have demonstrated that use of nano-size electrodes help in minimizing insertion induced stresses due to facile strain relaxation [2]. The theoretical framework has been earlier described by several researchers [3-6] but these models have been used to study intercalation induced stresses for materials whose volume change during lithiation is minimal (~6-8%) compared to high capacity materials such as silicon (~300%).

In this talk, development of a transient 1D axisymmetric model to simulate lithium insertion in different type of nanostructured Si electrodes will be presented. The physics of the model includes lithium transport, chemical and elastic strains, Li transport inducted stresses, and volume associated expansion. Case studies on constant current lithiation of different nanostructures, e.g. nanowires [2], nanotubes [7] and core-shell structures will be discussed [8]. Finally, a comparative study will be presented with respect to mechanical stability during lithium insertion among the different nanostructures.

Acknowledgement:

The financial support for this work provided by Applied Material Inc. is acknowledged.

References:

1. V. L. Chevrier and J. R. Dahn, J. Electrochem. Soc., 156, A454 (2009)

2. C. K. Chan, H. L. Peng, G. Liu, K. Mcllwrath, X.F. Zhang, R. A. Huggins, and Y. Cui, Nat. Nanotechnol., 31 (2008)

3. J. Christensen and J. Newman, J. Electrochem. Soc., 153, A1019 (2006)

4. J. Christensen and J. Newman, J. Solid State Electrochem., 10, 293 (2006)

5. R. Deshpande, Y. T. Cheng, and M. W. Verbrugge, J. Power Sources, 195, 5081 (2010)

6. Y. F. Gao and M. Zhou, J. Appl. Phys., 109, 014310 (2011)

7. T. Song, J. L. Xia, J. H. Lee, D. H. Lee, M. S. Kwon, J. M. Choi, J. Wu, S. K. Doo, H. Chang, W. il Park, D. S. Zang, H. Kim, Y. G. Huang, K. C. Hwang, J. A. Rogers, and U. Paik, Nano Lett., 10, 1710 (2010)

8. W. Wang, R. Epur and P. N. Kumta, Electrochem. Commun., 13, 429 (2011)

(157g) Micro-Structural Optimization of MoO2-Based Anode for SOFC Applications

Kwon, B. W., Kyung Hee University
Marin-Flores, O., Washington State University
Norton, M. G., Washington State University
Ha, S., Washington State University
Originally presented on: 11/4/2013 17:27:00 - 17:49:00

Micro-structural optimization of MoO_2-based anode for SOFC applications

Byeong Wan Kwon[a], Oscar Marin-Flores[a,b], M. Grant Norton[b], Su Ha[a]

[a]*The Gene and Linda Voiland School of Chemical Engineering and Bioengineering, Washington State University, Pullman, WA 99164*

[b]*School of Mechanical and Materials Engineering, Washington State University Pullman, WA, 99164-2920*

Solid oxide fuel cells (SOFCs), which typically run at temperatures in the range of 600-1000Â°C, can be operated with a variety of fuels, such as liquid hydrocarbons and synthetic liquid bio-fuels. The SOFCs are composed of a solid dense electrolyte sandwiched by two electrodes, an anode and a cathode. These electrodes should have a proper porosity to allow an efficient transport of both the reactants and byproducts, while possessing a proper sintering to allow high ionic and electronic conductivities. Therefore, their optimized micro-structures can lead to improve overall performances of SOFCs.

Recently, our group has successfully fabricated a novel MoO_2-based anode for the SOFC application and operated it with both the model aviation fuel and bio-diesel. In this study, we have improved the overall performance of our MoO_2-based SOFC by controlling the micro-structure of the MoO_2-based anode using a pore former such as a polyvinyl alcohol (PVA). Typically, MoO_2-based anodes were manufactured using MoO_2 nanoparticles to forma seed layer structure with an average thickness of 8 μm via electrostatic spray deposition (ESD) method. Over this seed layer, additionally catalyst inks containing both MoO_2 and different concentrations of PVA were painted to determine the effect of PVA concentration on anode morphology and power density output. We have utilized n-dodecane as a model fuel to assess the performance of SOFCs. The power density value at 0.6V improves from 540mW cm^{-2} to 2,168mW cm^{-2} as the PVA concentration increases from 6wt% to 12 wt%. However, further increasing the PVA concentration up to 21wt% decreases the power density output. Hence, the PVA concentration greatly influences the cell performance and a 12wt% PVA seems to be the optimum concentration that leads to the best micro-structure of the MoO_2-based anode. As observed in SEM images of 24h tested cells, the anode structure with a low PVA content of 6wt% is much denser compared to the ones with a higher PVA content, which can be attributed to a deficit amount of PVA required to produce a proper micro-porous structure in the anode. Therefore, a low porosity in the anode significantly limits mass transfer process, which leads to a larger anode overpotential (i.e., a larger anode charge transfer resistance from the impedance measurements) and lower power density output. Hence, the PVA concentration plays an important role in the performance and morphology of MoO_2-based anode. The MoO_2-based SOFC with the optimized micro-pore structure also showed a very stable performance over 24h test without showing any coking. Under the similar operating conditions used for the MoO_2-based SOFC, the commercial Ni-based SOFC showed a much lower initial performance and deactivated immediately due to the anode coking.

(182a) Economic and Engineering Assessment of Combined Heat and Power Fuel Cell Systems Installed in Commercial Buildings

Brooks, K., Pacific Northwest National Laboratory
Pilli, S.,
Originally presented on: 11/4/2013 15:15:00 - 15:37:00

PNNL support of the DOE's Fuel Cells Technology Office is to identify near term niche markets where fuel cells have potential, work with the DOE and stakeholders to develop activities in those areas, analyze the business case, and present the results to the community. Accelerating fuel cell use in these key early markets will create jobs in an industry that needs high volume purchases to ramp up production, decrease costs, support commercialization, and enable a domestic supplier base. PNNL has contracted ClearEdge Power to demonstrate fifteen 5 kW combined heat and power fuel cells in a wide variety of business in California and Oregon. These fuel cells are being remotely monitored for such things as electrical and thermal energy, overall efficiency, and availability. The gathered information will be analyzed, to provide "real-world" data from units "in the customer's hands" to validate performance, durability and reliability; installation, operations, and maintenance costs; and identify remaining barriers to widespread commercialization. As part of this assessment, a business case has been developed estimating of the size of the market and its growth potential, identifying possible niche markets, and comparing the CHP FCS with its alternatives in terms of economics, engineering and environmental impact. The results of this assessment will be presented.

(182b) Systematic Approach for Energy Efficient Eco-Industrial Parks

Noureldin, M. B., Saudi Aramco
Qahtani, S. A., Saudi Aramco
Owaidah, M., Saudi Aramco
Nutaifi, A., Saudi Aramco
Originally presented on: 11/4/2013 15:37:00 - 15:59:00

Sustainability since its original conceptualization is considered to have four dimensions; social; economic; environmental and institutional. It is simply the improved management of natural resources within a business establishment to provide economic and social benefits to the business and its surroundings. Eco-industrial parks (EIP) can serve a significant role in realizing economic, environmental and social benefits both to individual plants and to network of companies. Even though EIP are often publicized as a mean of reducing environmental damage through reduced waste, they can have number of benefits at different levels. Currently the major driving factor for the creation of most of it is to attain financial gain.

Industrial ecology concept may promote a new path of local development through the transition from industrial clusters to EIPs via exploiting common features of both models. Implementing Eco-Industrial Parks' (EIPs) principles in an existing industrial cluster or in the planning of new ones, represents a significant opportunity for its revitalization. IEP could exploit synergies from industrial clusters and non-industrial activities to easily create new production models in which the economic and environmental dimensions are symbiotic.

The well planning of new EIP or transformation of conventional industrial complexes to a one can brings significant value to energy efficiency and energy-based GHG emissions reduction. This task is a huge multi-variable multi-dimensional optimization problem in which the total eco-industrial park network depends on as small as stream condition and as big as the whole park functionality. The customized integration among many industrial plants and non-industrial ones in adjacent geographical locations can bring in more degrees of freedom to the "waste energy recovery" optimization problem and consequently presents new horizon to the radical energy-based GHG emissions reduction to levels never thought of before. Fortunately, many of the reasons which were hindering the application of inter-systems integration are nowadays addressed in cost-effective ways. However, to date EIPs key success has been a sequence of independent economically driven actions. Such evolutionary pattern may not be easily transferred to others. A holistic approach in addressing the problem using systematic techniques in the implementing of necessary modifications in either the brownfields or in the planning of Greenfields can bring huge value to energy resources conservation and GHG emissions reduction and is desperately needed today. It is very difficult with the current methods and tools to magically manufacture EIPs to work from scratch. First, there should be the basic ingredients in place, namely the desire of plants to actively participate and second the correct EIP membership and structure of firms. These basic ingredients can then be enhanced and improved upon, with correct support structure in place. The key to developing a successful EIP is to determine the best connections among different industrial plants and its surrounding communities in an EIP. Specifically it is the material and energy flows relationship among the different members in the alliance of plants which permit establishing optimal linkage to form a fruitful inter-dynamic structure. If such structure does not exist a successful EIP cannot be realized.

This paper addresses the energy efficiency of EIP via new systematic approach. It is different in concept than the current approach in which only the "transfer of waste heat" from one firm to a nearby one is conducted after exhausting all the possibilities to optimize the stand-alone plant in the firm. The total site targeting based on the pinch technology, which is the most widely used to date, starts with intra-systems integration which means again that only the waste heat of one member in the EIP is shared with other parks' members. Certainly, such level of sharing philosophy is not the ideal proactive form of true cooperation.

This paper introduces new method that achieves ambitious levels of energy efficiency and GHG emissions reduction via identifying best matchings' among EIP members; generating technically viable energy efficient alternatives; and defining the best generation and allocation of energy utilities systems. It also presents two new mathematical programming and constraints logic propagation tools to systematically find the best and the second best and so on matchings' among several plants for optimal eco-industrial park membership forms under all possible combinations of plants-specific design and operations modifications. The paper also introduces new graphical tool that enables the EIP designer to gain insights and create initial solutions for optimal energy loads for exchange among the members to best enhance the energy efficiency and GHG emissions of the whole park ("One for all and all for one philosophy"). The advantages of the new method and tools are illustrated and demonstrated by two case studies.

(182c) The Potential Of Osmotic Membrane Dehumidification

Kesten, A. S., Nanocap Technologies LLC
McCutcheon, J. R., University of Connecticut
Girelli, A.,
Blechner, J. N., Nanocap Technologies LLC
Originally presented on: 11/4/2013 15:59:00 - 16:21:00

THE POTENTIAL OF OSMOTIC MEMBRANE DEHUMIDIFICATION

Arthur Kesten, Jeffrey McCutcheon, Ariel Girelli and Jack Blechner

An osmotic membrane dehumidifier can use a flexible, semi-permeable membrane to facilitate capillary condensation of water vapor and the transport of condensed water through the membrane into a salt solution by osmosis. Here a humid gas stream is brought into contact with a semi-permeable membrane, which separates the gas stream from an osmotic (e.g., salt) solution. Some of the pores of the membrane are small enough to permit capillary condensation. Liquid formed within these pores can connect with liquid formed in adjacent pores, collectively forming continuous paths of liquid. These ?liquid bridges' extend across the thickness of the semi-permeable membrane and provide paths by which water can travel across the membrane. Because the membrane is so thin, water concentration gradients across the membrane can be large. This can provide a large driving force for water transport between the humid air and the osmotic fluid. The flexibility of the polymeric membrane allows for considerable design flexibility that enhances the potential for retrofit with any cooling system. An illustration of this two-step spontaneous process is given below.

Liquid desiccants are particularly effective as osmotic agents because water entering the desiccant solution is bound to the salt as water of hydration.Â This enhances the water concentration gradient across the membrane.

Laboratory testing of membranes and draw solutions under different environmental conditions is conducted using a cell comprised of two halves separated by a membrane. On the top half, the humid air is passed above the membrane. The draw solution is pumped through the bottom half of the cell; Â the draw solution is contained in a reservoir that is placed on a digital scale that measures mass to a hundredth of a gram. Humid air contacting the membrane condenses by capillary condensation in the pores of the membrane and the condensed water is drawn by osmosis into the draw solution. Mass changes versus time are recorded with the digital scale that is capable of measuring a maximum of 4,000 grams.

Osmotic dehumidification begins with condensation of water molecules, followed by the osmotic draw of the liquid water out of the pores into a draw solution. If the rate-limiting step is the osmotic de-swelling of the membrane, the highest osmotic pressure should be used to maximize flux. However, if capillary condensation is the limiting step, a threshold is reached beyond which increased osmotic pressure does not increase flux. To understand these limits, we tested three concentrations of magnesium chloride with a cellulose acetate membrane under identical conditions. The water flux varied directly with concentration suggesting that the limiting step for this membrane was the osmotic removal of liquid water as opposed to capillary condensation.

A separate experiment was constructed to compare the effectiveness of no membrane to our capillary condensation/osmotic dehydration system. Having no membrane in the system is the equivalent of exposing liquid desiccant directly to a humid air stream. Measured water removal rates for that system were less than one fifth of the rates for the membrane/draw solution tested here. And, of course, with no membrane between the humid air and the desiccant, there is always the potential for entraining desiccant in the air stream.

Cooling of the osmotic solution results in more effective dehumidification. Lower water vapor pressure leads to capillary condensation at lower relative humidity. For modest levels of cooling, the relative humidity exiting the Nanocap process will be around 50%.

Scale up of the process to larger surface areas of membrane can be accomplished with plate and frame systems, spiral wound devices and bundles of cylindrical fibers. Results with a plate and frame system demonstrate the effectiveness of a mathematical model as a basis for design of a larger system.

OSMOTIC DEHYDRATION

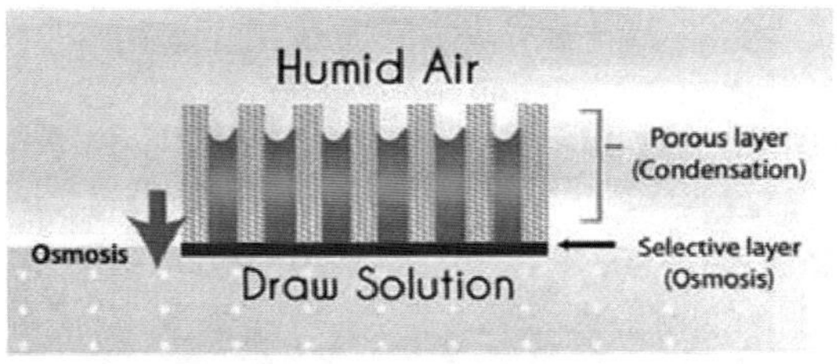

Osmotic dehydration across an anisotropic membrane consisting of a selective layer that is supported by a porous layer. Capillary condensation allows for osmosis to occur across the selective layer of the membrane into a concentrated salt or osmotic solution.

THE NANOCAP PROCESS

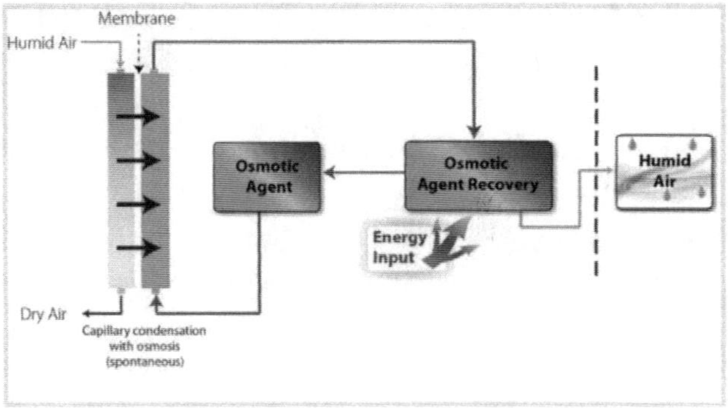

Energy used to regenerate the osmotic agent is intended to be a solar or waste heat source. Regeneration of the osmotic agent can be continuous or intermittent; the optimum time to regenerate occurs when solar or reject heat is available.

(182d) Improvement of Distillation Column Efficiency By Integration With Organic Rankine Power Generation Cycle

Sladkovskiy, D. A., St.Petersburg State Institute of Technology (technical university)
Originally presented on: 11/4/2013 16:21:00 - 16:43:00

Distillation as main separation technology consumes greater than 40% of the energy used by the chemical and petroleum process industries but the energy efficiency of a distillation column is low. It is well-known that significant energy savings can be achieved if complex column configurations and heat integration are used, although these improvements withdraw the attractive advantages of conventional distillation column such as flexibility, low capital investment and low operational risk.

This theoretical research focuses on improvement of distillation column vapor generation procedure by expansion turbine installation.

The liquid stream from the column bottom is pressurized by a pump and the high pressure liquid is vaporized in the reboiler. The resultant high pressure vapor stream drives the expansion turbine, which is coupled to the electricity generator or another rotating equipments. The turbine exhaust vapor enters the column bottom.

Such distillation column with power generation cycle (DCPGC) is kind of combination of conventional distillation column and organic Rankine cycle (ORC) power plant. The ORC is a cost efficient and proven method of converting low temperature waste heat to mechanical and/or electrical energy.

Such chemical components and its mixtures as propane, i-butane, n-butane, i-pentane et al. which occurs in distillation columns bottoms are known to be utilized or considered as ORC working fluids. So DCPGC is seem to be practically feasible.

DCPGC reboiler duty is about 10-20% greater than the conventional column reboiler duty but almost whole of extra heat transforms to electricity power. The theoretical DCPGC electricity generation efficiency is more than 90% (the heat required for the distillation remains the same as in conventional column and is not included in the efficiency calculation).

Only thermodynamic properties and thermal stability of column's bottom product impact on DCPGC feasibility, but not the temperature difference between top and bottom of the column. This is an advantage of suggested technology to compare with heat integrated column with heat pumps where temperature difference is quite important.

This theoretical research describes practical and economical feasibility of DCPGC for separating propane/propylene, i-butane/n-butane, i-pentane/n-pentane, butane-pentane, benzene/toluene, xylene mixtures. Crude oil and naphtha fractionation are also investigated. For instance DCPGC for i-butane/n-butane binary mixtures separation with 25 tonne/h feed flowrate can produce 1 MW of electricity power.

(182e) High Surface Area Zeolite Membrane for Efficient Bio-Fuel Production

LIU, W., Pacific Northwest National Laboratory

Originally presented on: 11/4/2013 16:43:00 - 17:05:00

Distillation and adsorption has been commonly used for respective enrichment and polishing of ethanol fuels in today's ethanol plants. In this presentation, development of thin zeolite/metal sheet membranes of surface area packing density comparable to polymeric sheets is discussed to enhance ethanol dehydration energy efficiency and lower capital costs. The mechanical structural integrity of this new membrane product concept is substantiated with stress modeling. A relatively simple fabrication process that renders its low-cost manufacturing is delineated with experimental results. Separation process studies are conducted along with the membrane product development, which include (i) performance tests with real-world feeds and (ii) process design and simulation of membrane separation integrated with whole plants. The present membrane is highly permeable to water that significant one-pass dehydration can be obtained within a short residence time (seconds). However, the membrane performance can be drastically affected by the feedstock. Proper pre-treatment is necessary to achieve stable separation performances with the real-world feed. Based on experimental results, a two-stage membrane separation system is proposed to conduct ethanol enrichment and purification by replacing respective distillation and adsorption processes. More than 90% energy saving is shown using the membrane separation. Due to high permeance and high membrane area packing density, the membrane module estimated for a 40-million gal/year ethanol plant can be fairly compact, about the size of plate heat exchangers. This membrane technology is promising for retrofitting of existing corn ethanol plants and for building of new cellulosic ethanol plants in the future as well.

(182f) Syngas Mass Transfer Analysis in a Hollow Fiber Reactor

Orgill, J. J., Brigham Young University
Lewis, R. S., Brigham Young University
Atiyeh, H. K., Oklahoma State University
Originally presented on: 11/4/2013 17:05:00 - 17:27:00

Biofuel production via fermentation is produced primarily by fermentation of simple sugars. Besides the sugar fermentation route, there exists a promising alternative process that uses syngas (CO, H_2, CO_2) produced from biomass as building blocks for biofuels. Although syngas fermentation has many benefits, there are several challenges that still need to be addressed in order for syngas fermentation to become a viable process for producing biofuels on a large scale. One challenge is mass transfer limitations due to low solubilities of syngas species. The hollow fiber reactor (HFR) is one type of reactor that has the potential for achieving appropriate mass transfer rates for biofuels production, but a better understanding of mass transfer limitations associated with all syngas species is critical for assessing syngas usage during fermentation.

Mass transfer coefficients for O_2 and CO_2 have been typically assessed for HFRs using dissolved gas probes. In limited cases, CO has been studied using myoglobin assays. Though there are many reports on mass transfer coefficients in HFRs, studies are lacking that compare mass transfer coefficients of syngas species, including comparisons obtained from different measurement methods. Such comparisons are important to verify the accuracy of the experimental methods and to provide an increased understanding of the competing mass transfer limitations associated with syngas species.

In this study, we measured the mass transfer coefficient for O_2 in a PDMS hollow fiber reactor as a function of Reynolds number (flow rate) using a dissolved O_2 probe. The data was fit to a model characterizing O_2 mass transfer and the model was used to predict the mass transfer coefficients for CO and H_2. In order to verify the model predictions for these latter species, a second measurement technique using GC analysis was developed to measure CO and H_2 mass transfer coefficients. The GC technique was validated by comparing the mass transfer coefficient of O_2 from both GC and dissolved O_2 probe analysis. The mass transfer coefficients of both H_2 and CO in a HFR will be compared and the impact of such coefficients on syngas utilization will be presented. This study will greatly enhance the knowledge necessary to assess syngas usage and mass transfer limitations in a HFR used for syngas fermentation.

(182g) An Innovative Methanol Synthesis Process Using Self-Heat Recuperation

Kansha, Y., The University of Tokyo
Ishizuka, M., The University of Tokyo
Tsutsumi, A., The University of Tokyo
Originally presented on: 11/4/2013 17:27:00 - 17:49:00

The demand for methanol will continue increasing, since methanol is attracting as a fuel for fuel cells, and is an intermediate raw material of hydrogen and dimethyl ether (DME) which are categorized green energy source. As well as other chemical processes, many methanol synthesis processes have reactant recycle system with product separation due to the low conversion ratio of the reactor. For this product separation from reactant recycle stream, a gas-liquid separator or distillation process has often been used. However, this gas-liquid separator and distillation processes are well known as an energy consuming process. Thus, many chemical engineers and investigators managed to produce a catalyst which achieves high conversion from reactant to production in the reactor for whole process optimization.

Authors have developed self-heat recuperation technology, in which not only the latent heat but also the sensible heat of the process stream can be circulated without any heat addition, leading to reduction of energy requirement in several chemical processes. This self-heat recuperation technology is suitable for thermal and separation processes and is applied to several chemical processes for energy saving as case studies.

In this research, we investigated the feasibility of applying self-heat recuperation technology to the methanol synthesis process and developed an innovative process for methanol synthesis process from the energy saving point of view. By installing the self-heat recuperation technology to the methanol synthesis process, the energy consumption of the process can be greatly reduced. Thus, it can be said that installation of self-heat recuperation technology is the other option of process design method to achieve high conversion in the reactor.

(192a) Fabrication of Porous Carbon Nanofibers With Adjustable Pore Sizes As Electrodes for Supercapacitors

Tran, C., Drexel University
Kalra, V., Drexel University
Originally presented on: 11/4/2013 15:15:00 - 15:45:00

We report a facile method for obtaining extremely high surface area and uniformly porous carbon nanofibers for supercapacitors. Blends of polyacrylonitrile and a sacrificial polymer at different compositions have been electrospun into non-woven nanofiber mats with diameters in the range of 200–400 nm. Electrospun nanofiber mats are then subjected to carbonization to obtain porous carbon nanofibers (CNFs) as polyacrylonitrile converts to carbon and the sacrificial polymer decomposes out creating intra-fiber pores. Resultant porous CNFs exhibit specific surface area of up to 1600 $m^2\ g^{-1}$ with a large fraction of mesopores (2–4 nm). No additional chemical or physical activation process was used. We demonstrate the tunability of the pore sizes within CNFs by varying the composition. The non-woven fiber mats of porous CNFs are studied as free-standing electrode materials for supercapacitors eliminating the need for polymeric binding agents. Electrochemical measurements showed large specific gravimetric and volumetric capacitances of up to 210 F g^{-1} and 60 F cm^{-3} in 1 M H_2SO_4 at a high cyclic voltammetry scan rate of 100 mV s^{-1} due to the large fraction of mesopores. These materials retain 75% performance at a large current density of 20 A g^{-1} indicating excellent power handling capability.

(192b) Enhanced Energy Storage By Tunable Electrolyte Confinement in Structure-Directed CNT Arrays

Nicol, K. T., Mainstream Engineering Corporation
Zastrow, D., Mainstream Engineering Corporation
Hill, J. J., Mainstream Engineering Corporation
Originally presented on: 11/4/2013 15:45:00 - 16:00:00

The demand for power and energy dense storage devices is growing rapidly. While electrochemical double layer capacitors (EDLCs) offer high power density, many believe that high energy density is also obtainable through precise tuning of the electrolyte, electrode and structure of the overall cell. Multiwalled carbon nanotubes (MWNTs) are a promising electrode material for EDLCs because of their high electrical conductivity, chemical stability and exceptionally high surface area. Until single walled carbon nanotubes can be produced as solely conductive structures, dense and low-diameter MWNTs offer the best properties with regard to carbon-based EDLC electrode materials. When fabricated in anodic aluminum oxide templates, a MWNT is grown within each nanopore, resulting in an ultra-high density aligned array. This method produces a carbon electrode material with a significantly increased active surface area over commercially available materials. The energy density and capacitance of these materials can be further improved by tuning the size of the electrolyte domain around each MWNT. This is accomplished through partial and selective removal of the template, thereby confining the electrolyte domain. In typical EDLCs, the diffuse part of the double layer does not contribute to charge storage. However, we have observed an apparent charge storage enhancement in domain-confined MWNT electrodes. We will also discuss the tunable approach with regard to optimizing energy storage independent of the choice of REDOX couple or solvent.

(192c) Multifunctional Nitrogen-Rich "Brick-and-Mortar" Carbon As High Performance Supercapacitor Electrodes and Oxygen Reduction Electrocatalysts

Yu, D.,
Wei, L., Nanyang Technological University
Chen, Y., Nanyang Technological University
Originally presented on: 11/4/2013 16:00:00 - 16:15:00

Novel porous carbon materials with excellent electronic, chemical and structural properties and high nitrogen content are desirable for many applications. Here, we show the design and synthesis of a new multifunctional porous carbon material with unique "point to plane" architecture through a simple but effective activation-free procedure. Porous carbon derived from a new type of copolymer poly (vinylidene chloride-co-acrylonitrile) serves as the nitrogen-rich "mortar". Reduced graphene oxide layers work as "bricks" with an aim to provide an open nanoscale scaffold and connect porous carbon, as well as modulate the ratio between mesopores and micropores. This new material has large surface area (957 m^2/g), high nitrogen content (6.6 at %), excellent conductivity (up to 5.1 S/cm), and favorable hierarchical meso- and microporosity. Benefiting from these intriguing features, this material shows ultrahigh specific capacitance at 361 F/g in aqueous electrolyte, far superior to that of many carbonaceous materials reported so far. The as-assembled asymmetric supercapacitors with the designed carbon material as negative electrodes and porous cobalt oxide nanorods as positive electrodes deliver a high-energy of 50.1 Wh/kg with a cell voltage of up to 1.6 V. Further, this material shows a superb electrocatalytic activity for oxygen reduction reaction in alkaline medium comparable to that of 20 wt% platinum-carbon electrodes, with much better methanol tolerant and long-term durability. We expect that this uniquely designed nitrogen-rich porous carbon material with its simple and scalable synthesis method will have great potentials for various applications in energy storage, energy conversion and catalysis.

(192d) Multiwalled Carbon Nanotubes With Tuned Surface Functionalities for Electrochemical Energy Storage

Wei, H., Lamar University
Gu, H., Lamar University
Guo, J., Lamar University
Wei, S., Lamar University
Guo, Z., Lamar University
Originally presented on: 11/4/2013 16:15:00 - 16:30:00

In the paper, surface functionalities effect on the electrochemical properties and energy storage properties of multiwalled nanotubes (MWNTs) were investigated via two different functionalized MWNTs, one modified with ether and carboxylic acid functional groups (denoted as MWNTs-EC), and the other with ester functional groups (denoted as MWNTs-E) as the supercapacitor materials. The functionalities onto the nanotubes were confirmed by Raman spectra and caused a minor enhancement in the band gap obtained from the Uv-vis spectra. Cyclic voltammetry results reveal that the capacitance of the MWNTs-EC, and MWNTs-E, was significantly increased by 41.5% and 123.2%, respectively, when compared to the as received MWNTs at lower scan rates. The energy density is also found to increase by 36.1 and 130.0% for the former and the latter, respectively, than that of the as received MWNTs-As in the galvanostatic charge/discharge measurement. The increases in both the capacitance and the energy density might arise from the enhanced surface wettability and larger pseudocapacitance contributions as confirmed by the electrochemical impedance spectroscopy. An equivalent circuit was proposed to disclose the nature of the capacitive behavior of the electrodes. The cyclic stability studies show an excellent long term cycling stability of the MWNTs, which retain 100.0% of the charge storage or discharge capacity even after 1000 charge-discharge cycles.

(192e) Three-Dimensional Core@Shell Nanostructured Array for Microscale Electrochemical Energy Storage

Mao, Y., Washington State University
Sun, X., Kansas State University
Originally presented on: 11/4/2013 16:30:00 - 16:45:00

Microelectromechanical systems (MEMS) based autonomous and biomedical devices play an ever increasing role in our society by monitoring the structural health of major civil works, the environmental presence of toxic or explosive compounds, and the physical health of human patients. Integration of robust and high power energy storage devices within these systems is important for reliable MEMS with necessary burst energy for data relay and dense components by reducing bulky external power sources. To overcome the shortcomings of conventional two-dimensional battery and supercapacitor configurations while developing more powerful energy storage devices, we focus on the development of three-dimensional architectures to take advantage of the vertical dimension-height. In this study, we designed a novel architecture composed of branched nanowire arrays, i.e. forest of nanotrees, to further boost electrochemical performance over one-dimensional nanowire arrays, inspired by branched trees with larger surface area to capture more sunlight. The prepared three-dimensional $ZnO@MnO_2$ core@shell branched nanowire arrays exhibit five times higher areal capacitance, better rate performance and smaller inner resistance than their nanowire array counterparts. These novel three-dimensional architectures offer promising designs for powering microelectronics and other autonomous devices on exceptionally small geometric scales.

(192f) Microfabricated Nickel Oxide Supercapacitors Based On High Aspect Ratio Concentric Cylindrical Electrodes

Armutlulu, A., Georgia Institute of Technology
Bidstrup Allen, S. A., Georgia Institute of Technology
Allen, M. G., Georgia Institute of Technology
Originally presented on: 11/4/2013 16:45:00 - 17:00:00

Supercapacitors are known as high-power energy storage devices, which narrow the application gap between the batteries and traditional electrostatic capacitors. As in the case of their electrostatic counterparts, they possess rapid charge and discharge capabilities with high degree of reversibility. Yet they can also store more energy per unit mass and/or volume which makes them superior to traditional capacitors and thus, yields a more attractive candidate for energy storage systems to be utilized in portable electronic devices. Supercapacitors can store energy electrostatically via separated charges in a Helmholtz double-layer that forms at the electrode/electrolyte interface, as well as electrochemically based on a phenomenon called pseudocapacitance that relies on reversible Faradaic charge transfer between the electrolyte and the surface of the electrode. In both energy storage mechanisms, electrode surface area plays a crucial role. In this study, our focus is on MEMS-enabled nickel oxide (NiO) based supercapacitors comprising high surface area 3D microelectrodes. The fabrication process consists of four major steps: formation of the array of high aspect ratio SU-8 pillars via backside exposure, deposition of alternating Cu and Ni layers on the SU-8 pillars by a sequential robotic electroplating process, mechanical and chemical etching of sacrificial Cu layers to obtain high surface area Ni structures in the form of concentric cylinders, and lastly conformal electrodeposition of the active material, i.e. nickel hydroxide ($Ni(OH)_2$), on the high surface area Ni backbone. Following the electrodeposition process of the $Ni(OH)_2$ layer, the structure is heated up to 300 ^{o}C in order to convert $Ni(OH)_2$ film to NiO, and the fabrication process is finalized. Performance tests for the resulting structures are carried out via cyclic voltammetry in 1 M KOH solution. Devices are cycled in excess of 500 times, and no significant capacitance degradation is observed. The areal and specific capacitance values of the supercapacitors are estimated to be 270 mF/cm^2 and 56 F/g, respectively.

(192g) Pretreatment Effects On Charge Storage of Early Transition-Metal Carbides and Nitrides

Djire, A., University of Michigan
Pande, P., University of Michigan
Sleightholme, A. E. S., University of Michigan
Deb, A., University of Michigan
Rasmussen, P. G., University of Michigan
Penner-Hahn, J., University of Michigan
Thompson, L. T., University of Michigan
Originally presented on: 11/4/2013 17:00:00 - 17:15:00

Pretreatment Effects on Charge Storage of Early Transition-Metal Carbides and Nitrides

Abdoulaye Djire[a], Priyanka Pande[a], Aniruddha Deb[b], Alice E. S. Sleightholme[a], Paul Rasmussen[ac], James Penner-Hahn[b], Levi T Thompson[ac]*

[a] Department of Chemical Engineering

[b] Department of Chemistry

[c] Hydrogen Energy Technology Laboratory

University of Michigan, Ann Arbor, MI_- 48109-2130

E-mail: ltt@umich.edu

Early transition-metal carbides and nitrides are promising candidates for use in supercapacitor electrodes due to their high electronic conductivities, high surface areas (can exceed 200 m^2/g), good electrochemical stabilities and high capacitance [1,2]. For example, the capacitance for VN has been reported to be as high as 1340 Fg^{-1} in aqueous KOH [3]. This high capacitance has been attributed to pseudocapacitive charge storage involving fast near-surface redox reactions. We observed that pretreatment of the surface to remove the passivation layer significantly improved the capacitances and electrochemical stabilities of early transition-metal carbides and nitrides in aqueous electrolytes. In this paper we present results from electrochemical characterization including cyclic voltammetry (CV), chronopotentiometry, and electrochemical impedance spectroscopy (EIS), and surface characterization using x-ray photoelectron spectroscopy (XPS) to explain changes in surface chemistry and electrochemical properties caused by pretreatment.Â

The early transition-metal carbides and nitrides of V, W and Mo were synthesized via temperature-programmed-reaction of their oxide precursors with 15% CH_4/H_2 or anhydrous NH_3 followed by passivation in 1% O_2/He at room temperature to form a oxygen-rich passivation layer preventing bulk oxidation on exposure to air [1]. Characterization of the structural properties was performed using nitrogen physisorption (BET surface area) and X-ray diffraction. Prior to the electrochemical characterization, the electrodes were pretreated in 0.3M NaOH aqueous solution for 1 min then rinsed with acetone, ethanol, and ultrapure water [4]. The CV was used to establish the stability windows and capacitances for these materials. The capacitance was deconvoluted into double-layer capacitance and pseudocapacitance by means of CV and EIS. Chronopotentiometry was used for further stability measurements and to determine changes in surface activities caused by pretreatment.

As shown in Figure 1, the specific capacitance for VN increased by 57% after pretreatment. This enhancement in performance suggests significant improvement in material electrochemical activities. We believe that pretreatment altered the charge storage for the carbides and nitrides. We also observed good electrochemical stability after successive cycling of the material for several thousand cycles. In general, the treated materials were more stable than the passivated materials.

Figure 2 shows the open circuit potential (OCP) or rest potentials for the passivated and treated VN in acidic medium. The OCP for the treated material is lower than that for the passivated material. We attributed this to changes in surface activities due to pretreatment.Â Using the Nernst equation, we found that the treated VN had a higher pseudocapacitance than the passivated VN. This pseudocapacitive charge-storage mechanism is being investigated using XPS and *in-situ* x-ray absorption spectroscopy.

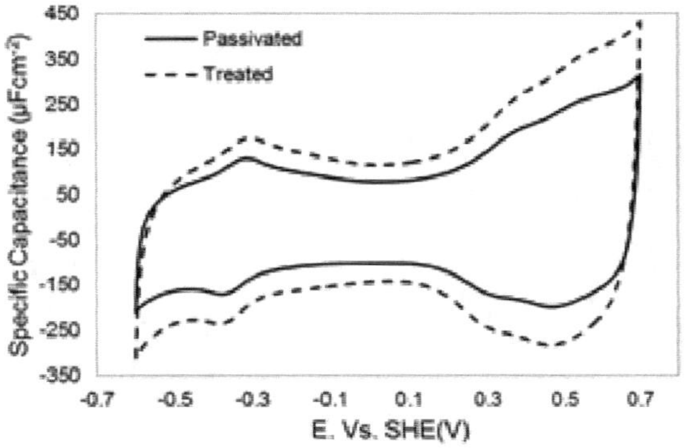

Figure 1: CV of passivated and treated VN in 0.1M H_2SO_4 at 50 mVs^{-1}.

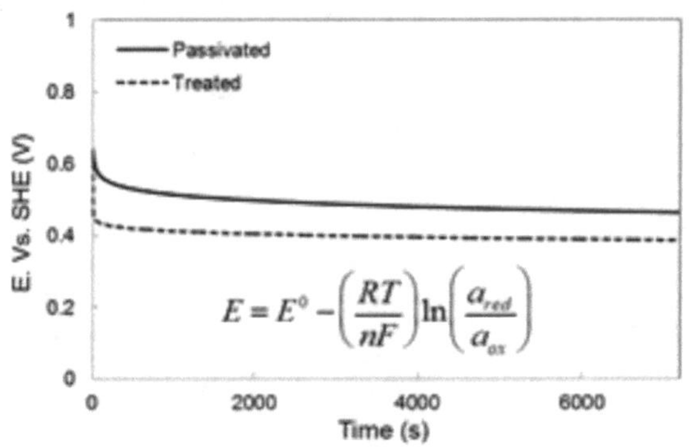

Figure 2: Rest potential of passivated and treated VN in 0.1M H_2SO_4

The physical and electrochemical properties will be described in terms of a proposed mechanism for psuedocapacitive charge storage.

References

(1) Cladridge, J. B.; York, A. P. E.; Brungs, A. J.; Green Malcolm L. H.; *Chem. Mater.* **2000**, 12, 132.

(2) Wixom, M.R.; Tarnowski, D. J.; Parker, J. M.; Lee , J.Q.; Chen, P. ?L.; Song, I.; Thompson, L. T.; *Mat. Res. Soc. Symp. Proc.* **1998,** 496, 643.

(3) Choi, D.;Â Kumta, P. N.; *Electrochem. Solid-State Lett.* **2005,** 8, 8, A418.

(4) Weidman, M.C.; Elposito, D.V.; Hsu, I. J.; Chen J. G.; J. Electrochem Soc.**2010**, 157, 12, F179

(192h) Doped Transition Metal Oxide Composite Electrodes for Supercapacitor Applications

Jampani Hanumantha, P., University of Pittsburgh
Kadakia, K., University of Pittsburgh
Hong, D. H., University of Pittsburgh
Poston, J., National Energy Technology Laboratory, U.S. Department of Energy
Ayyakkannu, M., National Energy Technology Laboratory
Kumta, P. N., University of Pittsburgh
Originally presented on: 11/4/2013 17:15:00 - 17:30:00

There is a dire need to provide a sustainable solution to the growing global demand for energy. It is this incessant demand that is the driving force fueling various current and emerging technologies including lithium ion batteries, supercapacitors and fuel cells. Supercapacitors occupy a unique place among these energy storage devices on account of their high power density and long cyclability. Supercapacitors are electrochemical charge storage devices capable of delivering large bursts of charge over a very short period of time. It is due to this unique ability that supercapacitors assume a very important role in the growing area of alternative energy storage systems. High surface area carbons, hydrated ruthenium oxide and MnO_2 are among the primary materials being considered for supercapacitor applications. Bulk oxide materials are traditionally hamstrung on account of their limited electrical conductivity for various electronic and electrochemical applications. In order to gain a fundamental understanding of the charge storage mechanisms in oxide materials, we deposited films of undoped transition metal oxides and doped transition metal oxides on current collectors and used various materials and electrochemical characterization techniques including X-ray photo-electron spectroscopy (XPS), cyclic voltammetry and electrochemical impedance spectroscopy to obtain a first-hand understanding of the underlying mechanisms. Effect of coating the various oxides on vertically aligned carbon nanotubes was also examined. Results of these studies will be presented and discussed.

(192i) Novel 3-D MnO2/Holey Graphene Nanostructure for Supercapacitor Applications With Enhanced Electrochemical Performances

Wang, L., Wayne State University
Meng, T., Wayne State University
Ara, M., Wayne State University
Deng, D., Wayne State University
Ng, S., Wayne State University
Originally presented on: 11/4/2013 17:30:00 - 17:45:00

As an excellent energy storage device with high-power density and long life cycle, supercapacitor has been intensely investigated in many aspects, such as electrode materials, electrode structure engineering, electrolyte compositions, separator improvements, as well as device design. Among them, one of the most urgent shortcuts is the around 10 times lower energy density compared to the traditional Li-ion batteries. This drawback has dramatically confined the applications of supercapacitors in electrical vehicles, intermittent energy storages and other occasions where both the power density and energy density are required. However, many researchers have been working on improving the energy density without significantly scarifying power density. These researches include increasing the voltage windows by employing organic electrolyte with larger voltage openings, applications of thinner and yet still mechanically strong separators, integrate materials with higher intrinsic capacitance, and so on. Despite all the effort on supercapacitor research, the current state-of-art on energy density still needs to be improved to meet the fast-growing application demanding in the emerging and potential applications. Therefore, we propose a mild method of preparing a novel 3-D MnO2/holey graphene composite for supercapacitor electrodes. The nano MnO2 nanorods in the composites serves as a capacitance booster due to its high intrinct capacitance and spacer to prevent 2-D graphene sheets from stacking back together. At the same time, graphene also serves both as active material to provide large surface area for charge storage and conductive agent to collect electrons generated by the surface redox reactions of semiconductive MnO2. Holes on graphene sheets enhance the efficient accessibility of electrolyte to the surface of active materials. With this novel 3-D composite, an increasement of around 30% to 50% in capacitance is expected.

(192j) Electrochemical Energy Storage of Magnetic Carbon Nanocomposites: Role of Magnetocapacitance and Magnetohydrodynamics

Zhu, J., Lamar University
Wei, S., Lamar University
Guo, Z., Lamar University
Originally presented on: 11/4/2013 17:45:00 - 18:00:00

Magnetic field has been used in different chemical processes to improve the quality of electrodeposited films, yield of photoredox products, and oxidative reactivity of catalysts and *etc*. In this work, a small external magnetic field will be imposed on a supercapacitor unit trying to improve its electrochemical energy density. Magnetic microtubular carbon nanocomposite fabrics decorated with uniformly distributed metal oxides (iron, cobalt and nickel oxides) nanoparticles will be fabricated via both microwave-assisted and conventional annealing processes. These fabrics could serve as flexible electrodes for supercapacitors to deliver high energy density and exhibit excellent cycling stability. The effect of metal oxide species, annealing method and applied magnetic field on the capacitance of each nanocomposite material will be tested using cyclic voltammetry and galvanostatic charge/discharge measurements. The corresponding mechanism will be investigated using electrochemical impedance spectroscopy. Cycling stability of each composite with and without magnetic field will be also comparatively studied.

(192k) Synthesis and Characterization of Orange Peel As Electrodes in Li-Ion Capacitors
Arie, A. A., Parahyangan University
Originally presented on: 11/4/2013 18:00:00 - 18:15:00

Orange peel was utilized as the raw material for activated carbon-based electrodes in Li-ion capacitors. The carbon samples weres synthesized
using a chemical activation process followed by a surface modification method. The surface modification method showed no significant effect on the specific surface area, however it changed the surface property of the carbon samples. Experimental results demonstrated that the electrochemical characteristics of un-treated activated carbons had a better specific capacitance compared with those obtained by surface modification method. Based on our results, orange peel waste can potentially be used as a cheap raw material for the mass production of activated carbons as electrodes in Li-ion capacitors.

(229a) A Reduced Order Lithium ION Battery MODEL Based On Smoothed Unit STEP Functions

Painter, R., Tennessee State University
Staples, C., Tennessee State University
Sharpe, L., Tennessee State University
Hargrove, S. K., Tennesee State University
Originally presented on: 11/5/2013 8:30:00 - 8:52:00

Abstract

Rigorous, physics based models for lithium ion batteries (LIB) have been developed over the past twenty years, beginning with Doyle et al. and Fuller et al. and continued by others [1-2]. These models are characterized by a large number of simultaneous partial differential equations (PDE) that require the solution of very large systems of non-linear algebraic differential equations (ADE). Problems that arise with the non-linear nature of the Butler-Volmer kinetics equation are perhaps even more significant than the computational time requirements for large systems. There is also a limited understanding of some of the physical parameters for lithium batteries that is compounded by variations in battery design and mass manufacturing processes. These uncertainties and the highly non-linear nature of the rigorous model pose problems for simulation and especially for control applications. These issues have led several researchers to develop reduced order models (ROM) based on various simplifications to the rigorous model. This paper presents a ROM that exploits the fact that the governing PDE's for the battery's positive electrode, separator and negative electrode differ only by the magnitude of certain physical parameters. Smoothed "If Then" unit functions are incorporated in a finite element LIB model at the battery's interfaces to switch physical parameters without injecting discontinuities into the model.

Key Words: Lithium Ion Battery, Reduced Order Model, Fuzzy If Then Logic

(229b) Performance Characterization of Electron-Beam Lithography Patterned Nafion Membranes for Direct Methanol Fuel Cells

Omosebi, A., New Jersey Institute of Technology
Besser, R. S., Stevens Institute of Technology
Originally presented on: 11/5/2013 8:52:00 - 9:14:00

Direct methanol fuel cells (DMFC) are capable of directly converting methanol into DC electricity without the need for a reforming step. As a liquid fuel, methanol is easy to store and transport. However, high catalyst loading requirements and performance losses continue to limit the wide-scale adoption of the DMFC. The loss in performance results primarily from anode-to-cathode methanol crossover and the slow reaction kinetics of both methanol oxidation and oxygen reduction at the anode and cathode respectively. New materials like sulfonated poly ether ether ketone (SPEEK) and sulfonated poly sulfone (SPSf), and surface treated Nafion membranes which exhibit lower methanol crossover are being investigated as replacements to pristine Nafion for the construction of the DMFC. At present, these innovations have resulted in a tradeoff between methanol permeation and ionic conductivity which consequently lead to inefficient use of the expensive metals used to form the catalyst layers. In order to maximize the utilization of the loaded catalyst, the reaction zone should be expanded without increasing transport resistance for reactants accessing the reaction plane. To achieve this objective, the membrane-electrode interface could be restructured to possess a larger interfacial area by creating features on the Nafion membrane. We present the use of electron beam lithography coupled with dry etching strategies for patterning Nafion and creating nano/micro-structured membrane electrode assemblies (MEA). EBL provides a convenient tool for rapid prototyping and assessing feature sizes beyond the resolution of conventional light-based lithographic techniques. In this communication we present the performance characterization of the electron beam patterned membranes for DMFCs.

(229c) A Simple Analytical Model of a Regenerative Hydrogen Bromine Fuel Cell

Martino, D., WPI
Datta, R., Worcester Polytechnic Institute
Originally presented on: 11/5/2013 9:14:00 - 9:36:00

A Simple Analytical Model of a Regenerative Hydrogen Bromine Fuel Cell

Due to the intermittent nature of renewable power sources, especially wind and solar, and the changing daily requirements of a power grid, energy storage solutions are needed to supplement the primary power sources. One promising option is the so called regenerative fuel cell (RFC). These are reversible fuel cells which store power by performing electrolysis to generate pure components which can then be reacted as fuels in a fuel cell to release the stored power. Hydrogen/oxygen polymer electrolyte membrane (PEM) systems however, suffer from poor performance when operated as a RFC, because of the sluggish oxygen reduction reaction (ORR). One potential way to improve performance on this type of cell is to use hydrogen/bromine PEM-based RFC, because of the much higher reactivity of bromine. A significant advantage of using the hydrogen bromide chemistry, thus, is that the performance of this system is limited by the transport and conductivity properties of the membrane, rather than by electrode kinetics.

To accurately design such RFC systems, it is useful to have a simple analytical model that can accurately describe the electrochemical behavior of a cell based on this chemistry. In this work, we propose a simple model for the H_2/Br_2 reversible fuel cell based on Butler-Volmer kinetics at the electrodes and detailed thermodynamic and transport considerations. The model can accurately predict cell behavior at different operating conditions, such as temperature, fuel concentration, and flowrate, while simplifying and reducing the number of fitted parameters required. An accurate and easily implemented model for a hydrogen bromine fuel cell enables better selection of cell materials and identification of more effective operational modes for power generation and storage.

(229d) Importance of Ion-Ion Interactions in Membranes for All-Vanadium Redox Flow Batteries

Griffith, L., University of Michigan

Kim, S. U., University of Michigan

Chen, H. Y.,

Monroe, C. W., University of Michigan

Originally presented on: 11/5/2013 9:36:00 - 9:58:00

Redox Flow Batteries (RFB) have the potential to provide the high capacity energy storage needed to improve power grid efficiency through peak shaving, as well as aid in the integration of renewable energy sources, such as wind energy and solar energy. The high cost of traditional ion selective membranes has motivated the exploration of the many different membranes available for use in RFBs[1]. In order to facilitate this exploration it is necessary to accurately characterize multi-component mass transport in various membranes, so the impact of using different membranes on RFB performance can be modeled.

Treatment of binary diffusion of species present in all-vanadium RFBs is ubiquitous in the literature[2], however, for these measurements to be useful in accurate mass transport models, the effects of ion-ion interactions on multicomponent transport across ion-selective membranes must be negligible. Heintz *et al*[3] have investigated these effects for sodium and hydrogen diffusing across neosepta membranes and have seen that the magnitude of the ion-ion interactions is comparable with the magnitude of the ion-solvent interactions. This shows that the importance of including ion-ion interactions in models of RFBs should be tested, so we have measured ion-ion interactions for $VOSO_4$ and H_2SO_4 in porous, cation-exchange, and anion-exchange membranes.

This talk will briefly discuss how Barnes's transient diffusion model[4] can be used to quickly measure binary diffusion coefficients and sorption equilibria. These results will then be used to measure ion-ion interactions through application of the Stefan-Maxwell formalism given in Equation 1.

$$-c_i \nabla \mu_i = \sum_{j \neq i} \frac{RT c_i c_j}{c_T D_{ij}} \left(\vec{v}_i - \vec{v}_j \right) \qquad (1)$$

For n diffusing species, this system of n force explicit equations can be reduced to a system of n-1 flux explicit equations by taking the membrane velocity as the reference and applying the Gibbs-Duhem equation[5]. Taylor and Krishna[6] show that a batch dialysis cell will decay exponentially toward equilibrium at long times, with decay constants that are the eigenvalues of the system of transport equations. Therefore, if the binary diffusion interactions are known, the ion-ion interactions can be measured with an interdiffusion experiment in a batch dialysis cell. The values of the binary diffusion coefficients as well as the ion-ion interactions will be given for a system composed of $VOSO_4$ and H_2SO_4 diffusing through porous, cation exchange, or an anion exchange membranes. The importance of ion-ion interactions in each membrane class will be discussed.

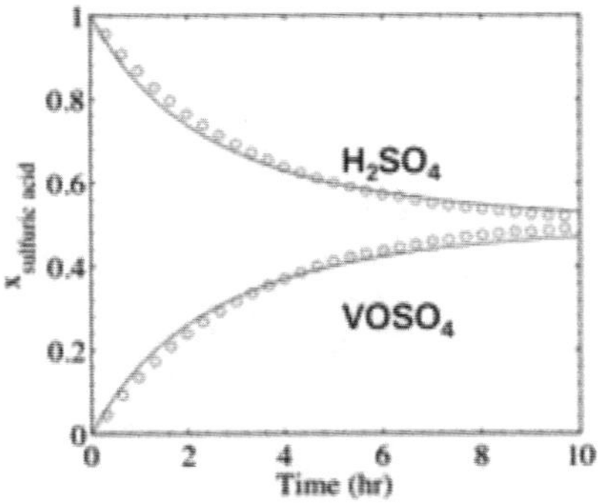

Figure 1: Experimental data for interdiffusion of $VOSO_4$ and H_2SO_4 across a Celgard membrane (o) fit using the ion-ion interactions as a parameter (-).

1. Skyllas-Kazacos, M., M. H. Chakrabarti, et al. (2011). J. Eletrochem Soc **158**(8): R55-R79
2. Wiedemann, E., A. Heintz, et al. (1998). J. Membrane Sci. **141**(2): 215-221.
3. Heintz, A., E. Wiedemann, et al. (1997). J. Membrane Sci. **137**(1-2): 121-132.
4. Barnes. (1934). Physics **5**(1): 5.
5. Meyers, J. P. and J. Newman (2002). J. Electrochem. Soc. **149**(6): A718-A728.
6. Taylor, R. and R. Krishna (1993). Multicomponent mass transfer, Wiley.

(229e) Mechanistic Studies of Nonaqueous Li-O2 Electrochemistry

McCloskey, B. D., IBM Almaden Research Center

Originally presented on: 11/5/2013 9:58:00 - 10:20:00

This presentation will highlight mechanistic studies of electrochemistry occurring at the cathode of a Li-O_2 battery. Quantitative Differential Electrochemical Mass Spectrometry (DEMS) and electrochemical measurements on flat, nonporous glassy carbon electrodes were coupled with *ex-situ* chemical analysis of electrodes to better understand the active chemistry within the cell. Our results indicate that solid Li_2O_2 formation and oxidation are the only active reversible electrochemical reactions occurring at the cathode and that both of these processes occur at low kinetic overpotentials, particularly compared to aqueous oxygen reduction and evolution. Furthermore, no solution-soluble intermediates, such as O_2^-, exist in a reasonably stable solvent (such as dimethoxyethane). Given the low overpotentials observed in the cell and the limited diffusion rates that are implied from no intermediate solubility, heterogeneous catalysis (beyond the catalytic effect observed from carbon) for oxygen reduction and evolution in the Li-O_2 cell may be unnecessary.

(229f) Confined PFSA-Zeolite Proton Conducting Membrane: The Effects of Confinement and Zeolite-Type

Sim, V., The Hong Kong University of Science and Technology
Poon, H. Y., The Hong Kong University of Science and Technology
Han, W., The Hong Kong University of Science and Technology
Budihardjo, F. F., The Hong Kong University of Science and Technology
Yeung, K. L., The Hong Kong University of Science and Technology
Originally presented on: 11/5/2013 10:20:00 - 10:42:00

Perfluorosulphonic acid (PFSA) membranes are a well-known proton exchange membrane and the 'gold standard' commercial Nafion membrane is renowned for its excellent proton conductivity and long term stability under fully hydrated conditions. However, their efficiency is highly dependent on the hydration level of the membrane, making water management an important issue. Traditionally, these membranes use external humidifying equipments to ensure optimum operation, which often translates to an increase in costs and lower efficiency. Thus, the PFSA membranes are highly sensitive to temperature, making operations at temperatures higher than 80°C difficult, although a higher temperature would increase the efficiency of the electrocatalysts and improve its poison tolerance. A proton conducting membrane that can operate effectively without humidifying aid and at higher temperature operations is very attractive.

Recently our lab reported a novel self-humidifying membrane that is obtained by confining Nafion PFSA within zeolite-coated pores of a stainless steel foil that displays superior performances compared to the commercial Nafion 117 membrane. A multi-layered zeolite was deposited on the stainless steel; a nonporous silicalite-1 zeolite layer served as an insulating and anti-corrosion coating, a thin layer of nanozeolites (including LTA, BEA, FAU, and MFI) for water retention and a surface layer of HZSM-5 zeolite as a protective layer. This work investigates the effect of confinement on the physicochemical and thermomechanical properties of PFSA. We also explore the effect of water retention properties of various zeolites on the performance of the PEMFC.

(229g) CFD Based Design of Planar SOFC for Improvement in Performance

Dey, T., Indian Institute of Technology Bombay
Singdeo, D., Indian Institute of Technology Bombay
Ghosh, P. C., Indian Institute of Technology Bombay
Bose, M., Indian Institute of Technology Bombay
Basu, R. N., Central Glass and Ceramic Research Institute, CSIR
Originally presented on: 11/5/2013 10:42:00 - 11:04:00

Solid oxide fuel cell (SOFC) is one of the possible solutions to the challenge posed by our current energy requirement, the core competency of which is based on efficiency and clean energy conversion. In order to achieve wide spread acceptance, the technology has to perform reliably well over a long period of time. Long term performance is governed by materials of construction and design of the cell as well as the environment under which it is operated. SOFC performances are affected by different types of losses such as activation, ohmic and concentration. The concentration losses depend upon the distribution of fuel and oxidant over the cell active area. The flow distribution in any fuel cell is ensured by designing appropriate flow-configuration. The present work aims to investigate the flow uniformity and mass transport losses in SOFC and the role of flow geometry on improving performance of the unit. The preliminary design consists of an outlet manifold, channel feeding the active area region and an inlet manifold. The simulation of the entire model on the computational platform is proved to be counter-productive. So, the geometry is subdivided into three parts (inlet manifold, straight channel and outlet manifold) and the flow field is sequentially simulated. Pressure drop in the channel is determined assuming the flow to be fully developed and laminar in the channel, which is valid owing to the large length to hydraulic diameter ($l/D >> 1$) ratio for each channel. The simulation in the other two decomposed geometries is obtained from numerical simulation. Different types of manifold modifications are used in order to achieve better flow uniformity. Finally, the design which results in to the most uniform flow distribution in the active area is selected for fabrication and performance testing. The electrochemical performance is observed to enhance by ~15% using the modified design of the manifolds.

(283a) Modeling Hydrogen Transport Into Palladium Hydride With Detailed Surface Reaction Chemistry

Salloum, M.,
James, S. C.,
Robinson, D. B., Sandia National Laboratories
Originally presented on: 11/5/2013 12:30:00 - 12:49:00

Palladium is an important hydrogen storage and separation material that offers high volumetric power density in metal hydride batteries and fuel cells. We present a multiphysics mathematical model of the coupled reactive flow of hydrogen isotopes in a packed palladium hydride bed. An isotope exchange experiment allows study of hydrogen transport independently of the lattice expansion from palladium to palladium hydride. The model considers rates of chemical reactions and mass transport within a hydride bed and incorporates the multistep reaction mechanisms that take place at the metal bulk, surface and gas levels. A novel feature in the chemical reaction model is that it allows accounting for all absorption, adsorption and diffusion activation energies that can be adjusted depending on the state of the metal surface. The simulation results are comparable to measured values. The model is useful in designing an optimal hydride bed operating at a given temperature, hydride particle size and surface condition. Based on the same chemical kinetic model, we also present a gas chromatography model of the exchange process. This model offers a better understanding of the exchange kinetics by exploring the specifics of the flow.

Sandia National Laboratories is a multi-program laboratory managed and operated by Sandia Corporation, a wholly owned subsidiary of Lockheed Martin Corporation, for the U.S. Department of Energy's National Nuclear Security Administration under contract DE-AC04-94AL85000.

(283b) Catalytic Aqueous Phase Reforming of Pine Wood, Switchgrass and Wastepaper for H2 and Biocrude Production

Tungal, R., South Dakota School of Mines & Technology
Ohlsen, K.,
Shende, R., South Dakota School of Mines and Technology
Originally presented on: 11/5/2013 12:49:00 - 13:08:00

The authors report (a) the subcritical aqueous phase reforming of lignocellulosic biomass (pine wood and switchgrass) using homogeneous $Ni(NO_3)_2$ and heterogeneous catalysts for the production of H_2 and biocrude/bio-oil, (b) synthesis of novel nanoscale catalytic materials and (c) detailed investigation of biocrude generated from a biomass. Different kinds of bifunctional catalytic materials were prepared using wet-impregnation, solid state synthesis and sol-gel approaches. In the case of sol-gel approach, metal (Ni, Fe)-salt was dispersed in ethanol/water solution containing surfactants such as Brij-76/pluronic 123 and gelation was achieved using propylene oxide. As prepared material was aged, dried, calcined at 400-600 ^oC. Catalyst nanoparticles were also synthesized using solid state synthesis method. The catalytic materials were characterized using X-ray diffraction, FTIR, SEM, TEM and BET. A biomass was loaded in the high temperature high pressure SS316 reactor and aqueous phase reforming was performed at 200 ^{o} - 300 ^oC. During reforming reaction, gas samples were withdrawn at regular time intervals and analyzed using GC equipped with thermal conductivity detector and Chrompack GC column. The product gas was found to contain mainly H_2, CH_4, CO and CO_2. The processed slurry was centrifuged to recover and biocrude and residue. The biocrude was further analyzed by TOC analyzer, GC-MS and HPLC. The results obtained on H_2 and biocrude composition obtained during aqueous phase reforming of biomass substrates will be presented.

(283c) A Computational Study of the NaBH4 Hydrolysis Reaction for Hydrogen Production

Li, P., University of Pittsburgh
Henkelman, G., University of Texas at Austin
Johnson, J. K., University of Pittsburgh

Originally presented on: 11/5/2013 13:08:00 - 13:27:00

$NaBH_4$ is a promising hydrogen storage material due to its high hydrogen content (10.7 wt%), relative safety, and low cost. The $NaBH_4$ hydrolysis reaction: $NaBH_4 + 4H_2O \rightarrow NaBO_2{\cdot}2H_2O + 4H_2$ has found application in niche markets such as portable energy generation. The rate of hydrolysis reaction without catalyst is very low in aqueous solution, so a great deal of work has been focused on catalyst design and application.[1,2] Additionally, the aqueous hydrolysis reaction also suffers from low system hydrogen storage capacity due to excess water used as solvent for $NaBH_4$ and byproducts. Recently, Matthews and coworkers demonstrated that steam hydrolysis of $NaBH_4$ can be used to release hydrogen with both fast kinetics and high extent of reaction without the use of any catalyst and with reduced water consumption.[3-5] Despite a long history of research and development, the basic mechanism of $NaBH_4$ hydrolysis reaction is not understood on a molecular level. We have performed first-principles density functional theory calculations to investigate the elementary reaction steps in aqueous $NaBH_4$ hydrolysis. Detailed understanding of these steps would be useful for reaction condition optimization, catalyst design, and ultimately for precise control of the reaction in practical applications. Representative minimum energy pathways have been characterized from the family of mechanisms that exist in the real system by using the conventional nudged elastic band method[6] and the solid-state nudged elastic band method.[7] We identified energy barriers and reaction intermediates for each elementary step. Taken together, these steps provide a full picture of the $NaBH_4$hydrolysis reaction at the molecular level.

References

(1) Schlesinger, H. I.; Brown, H. C.; Finholt, A. E.; Gilbreath, J. R.; Hoekstra, H. R.; Hyde, E. K. *Journal of the American Chemical Society* **1953**, *75*, 215-219.

(2) Liu, B. H.; Li, Z. P.; Suda, S. *Journal of Alloys and Compounds* **2006**, *415*, 288-293.

(3) Aiello, R.; Sharp, J. H.; Matthews, M. A. *International Journal of Hydrogen Energy* **1999**, *24*, 1123-1130.

(4) Marrero-Alfonso, E. Y.; Gray, J. R.; Davis, T. A.; Matthews, M. A. *International Journal of Hydrogen Energy* **2007**, *32*, 4717-4722.

(5) Marrero-Alfonso, E. Y.; Gray, J. R.; Davis, T. A.; Matthews, M. A. *International Journal of Hydrogen Energy* **2007**, *32*, 4723-4730.

(6) Henkelman, G.; Uberuaga, B. P.; Jonsson, H. *J. Chem. Phys.* **2000**, *113*, 9901-9904.

(7) Sheppard, D.; Xiao, P.; Chemelewski, W.; Johnson, D. D.; Henkelman, G. *J. Chem. Phys.* **2012**, *136*, 074103-074108.

(283d) Metal Organic Framework for Hydrogen-Storage Applications

Sahu, D.,
Verma, A., IIT Guwahati
Gumma, S., Indian Institute of Technology Guwahati
Originally presented on: 11/5/2013 13:27:00 - 13:46:00

Fuel cell run vehicles sooner or later will be a reality, needs an on-board storage that allows driving range in excess of 400 km, thus requires a tank containing 4 kg of H_2. Since hydrogen is a light weight gas, significant amount of it can be stored in practical size of container only by lowering the temperature to cryogenic temperature (less than 100 K). Thus storage tank filled with solid adsorbents like carbons, zeolites, Metal organic frameworks (MOFs) etc can be a better choice to store same amount of hydrogen but at relatively elevated temperature.

In this work, we report the hydrogen adsorption properties of four different MOFs namely Zn-BDC, Cr-BDC, Ni-DABCO and Mg-DOBDC. Gravimetric hydrogen adsorption measurements were performed for a wide range of temperature (90 - 298 K) and pressure (0 - 100 bar). At 90-100 K and ~100 bar, all the adsorbents attain saturation which indicates pore filling of adsorbents at this temperature and pressure. On the other hand at room temperature, isotherms are in Henry's region and metal site unsaturation governs the adsorption. Net adsorption isotherms are obtained because excess or absolute adsorption isotherms do not account the loss of tank volume by adsorbent solid itself. Tank volume calculation shows that the volume required will be ~20 - 30% less with Mg-DOBDC as adsorbent than with empty tank to store similar amount of hydrogen at 100 bar.

(283e) First Principles Investigation of Ternary and Higher Metal Hydrides for High Temperature Tritium Sequestration Application

Nicholson, K., Georgia Institute of Technology
Sholl, D., Georgia Institute of Technology
Originally presented on: 11/5/2013 13:46:00 - 14:05:00

The U.S. Department of Energy Next Generation Nuclear Plant (NGNP) demonstration project will use a very high temperature gas-cooled reactor with helium coolant outlet temperatures in the range of 1000-1200 K. To avoid quenching this potentially industrially-useful process heat, we apply first principles methods to identify metal hydrides capable of removing the dilute tritium contaminant from the high temperature coolant. The thermodynamics of most binary metal hydrides (one metal constituent) are known, and ternary (or higher) systems are typically less thermodynamically stable than the associated binary hydrides, with exceptions for the Th-Zr-H system and complex transition metal hydrides. We first investigate the anomalous ternary interstitial Th-Zr-H system to corroborate the experimentally-observed enhanced thermodynamic stability as a proof of principle calculation for this class of materials. We compute phase diagrams using a grand potential minimization technique and account for configurational entropy effects using a cluster analysis method. We then screen a library of more than 100 complex transition metal hydrides to identify materials with enhanced thermodynamic stability relative to the associated binary hydrides and to rank candidate materials for the NGNP application.

(283f) Evaluation of the EOR Potential in Shale Oil Reservoirs By Water and Gas Flooding

Chen, K., Texas Tech University
Sheng, J.,
Originally presented on: 11/5/2013 14:05:00 - 14:24:00

Evaluation of the EOR potential in Shale Oil Reservoirs by Water and Gas Flooding

ABSTRACT

With the relatively modest natural gas price, producing oil from unconventional shale reservoirs, which are less common and less well understood than conventional sandstone and carbonate reservoirs, has attracted more and more interest from oil operators. Though many tremendous efforts have been done on the development of shale resources, the ultimate oil recovery is still low (5-10%). For the important role of shale resources in future oil and gas industry, more stimulation and production strategies must be considered and tested to find better methods to improve oil production from shale reservoirs.

In this paper we use the simulation approach to evaluate the EOR potential in shale oil reservoirs by gas flooding and water flooding. We first validate a base model, and discussed the determination of miscibility parameter and injection pressure. Production behavior and oil recovery of different schemes are discussed through sensitivity studies. Simulation results of primary production, gas injection and water injection are compared in this paper. Results show that miscible gas injection has better effect on improving oil recovery from shale oil reservoirs than water injection. The solvent injected into the reservoirs above MMP can be fully miscible with oil, and reducing oil viscosity greatly, in addition to the mechanism of pressure maintenance. Our simulation results indicate that the oil recovery can be increased up to 15.1% by using gas injection in a hydraulically fractured shale reservoir, compared with the original 6.5% recovery from the primary depletion. The oil recovery from water flooding is about 11.9%.

The results indicate that miscible gas flooding could be a good prospect in developing shale oil resources.

(283g) Component Development for the Reformer Steam Iron Process for Decentralised Sustainable Hydrogen Production

Nestl, S., Graz University of Technology
Voitic, G., Graz University of Technology
Geymayer, M., Graz University of Technology
Hacker, V., Graz University of Technology
Originally presented on: 11/5/2013 14:24:00 - 14:43:00

In a world with an increasing population and energy demand, fuel cells have the potential to realise high efficient and emission-free power generation based on renewable hydrogen. At present the majority of the worldwide hydrogen production is based on fossil fuels which are converted in centralised plants and delivered to a small number of large-scale consumers. The future energy system will be based on hydrogen produced onsite near the customer, based on renewable energy carriers. This avoids the emission of carbon dioxide and the long distance transport of hydrogen.

The steam reforming of biogas is one promising process for renewable decentralised hydrogen production. It can be seen as a combination of the conventional steam methane reforming reaction (1), the dry reforming reaction (2) and the water gas shift reaction (3).

$$CH_4 + H_2O \rightarrow CO + 3H_2 \qquad \Delta H_{298} = +206\ kJ \qquad (1)$$

$$CH_4 + CO_2 \rightarrow 2CO + 2H_2 \qquad \Delta H_{298} = +247\ kJ \qquad (2)$$

$$CO + H_2O \rightarrow CO_2 + H_2 \qquad \Delta H_{298} = -41\ kJ \qquad (3)$$

The conventional steam methane reformer requires additional purification steps in order to remove the produced carbon dioxide and unreacted carbon monoxide and methane. The steam iron process enables the production, purification and storage of hydrogen produced from any hydrocarbon containing feedstock in one single unit. At this chemical looping process a solid contact mass that consists of magnetite (Fe_3O_4) is reduced by hydrogen and carbon monoxide which is produced at the reforming process (4). The steam oxidation of the newly formed iron leads to the formation of pure hydrogen (5).

$$Fe_3O_4 + 4H_2/CO \rightarrow 3Fe^0 + 4H_2O/CO_2 \qquad (4)$$

$$3Fe^0 + 4H_2O \rightarrow Fe_3O_4 + 4H_2 \qquad (5)$$

The reaction conditions at the reformer stage have high influence on the reactivity and stability of the solid contact mass. In order to reach high reaction rates in the steam iron reactor the amount of carbon dioxide and steam in the syngas have to be minimised by working at low steam to CH_4 (S/C) ratios and high temperatures. On the other hand low S/C ratios reduces the CH_4conversion and may cause coke formation on the reforming catalyst and the contact mass which in turn reduces the purity of the produced hydrogen. By increasing the process temperature these effects can be minimised but particular attention has to be given to the stability of all components. In this work different process conditions are investigated in order to optimise the system.

Acknowledgment: This work is funded by the Research Studios Austria program of the *Austrian Federal Ministry of Economy, Family and Youth*.

(283h) Bio-Hydrogen Production By Mixed Bacteria From Anaerobic Sludge With Different Combined Pretreatment Methods

Sui, H., Tianjin University
Wu, M.,
Zhang, R., Tianjin University of Technology
Li, X., Tianjin University
Originally presented on: 11/5/2013 14:43:00 - 15:02:00

Microbial hydrogen production by anaerobic bacteria has been extensively investigated since hydrogen is considered to be a future clean source of energy. Dark fermentation systems offer an excellent potential for practical application with emerging hydrogen and fuel cell technologies. Many investigations focus their attention on the pure bacteria, and a number of known bacteria have been used to produce hydrogen gas[1,2,3,4]. But the process is limited by the distribution and immobilization of the bacteria. So there are also some studies on the anaerobic sludge which is used as the source of the mixed bacteria during the dark-fermentation. However, the hydrogen yield rate is low (Currently fermentative processes produce 0.48 to 1.63 mole of hydrogen per mole glucose)[5,6,7].

In order to enrich hydrogen producing bacteria and increase the hydrogen yield, four combined pretreatment methods including heat+ acid, heat+ alkali, acid+ heat and alkali+ heat were performed on the seed sludge in this paper. During the dark fermentation, the theoretical hydrogen yield is 4mol H_2/mol glucose when the process is acetic-acid type fermentation. Generally, 25-50% of the theoretical hydrogen yield can be obtained from sugar fermentation with mesophilic temperatures and moderate microorganisms[8,9,10,11]. Results showed that hydrogen yield increased greatly from 1.16 mol H_2/mol glucose (heated only) to 1.867 mol H_2/mol glucose through the optimization of the combined pretreatment methods. And the highest hydrogen concentration reached 85% at 80°C for 40min at pH3.0 (heat+acid).

Since the process of photo-fermentation can utilize the volatile fatty acids (VFAs) produced by the dark-fermentation and realize continuous and stable H_2 production. In this study, the pretreated bacteria were continually enriched with the production by the dark-fermentation. Then the enrichment cultures were carried out the photo-fermentation, the hydrogen yield could reach 6.24mol H_2/mol glucose.

Reference

[1] Zhang H, Bruns MA, Logan BE. Biological hydrogen production by Clostridium acetobutylicum in an unsaturated flow reactor[J]. Water Res, 2006(40):728–734.

[2] Ishikawa M, Yamamura S, Takamura Y, Sode K, Tamiya E, Tomiyama M. Development of a compact high-density microbial hydrogen reactor for portable bio-fuel cell system[J]. Hydrogen Energy, 2006(31):1484–1489.

[3] Turcot J, Bisaillon A, Hallenbeck PC. Hydrogen production by continuous cultures of Escherchia coli under different nutrient regimes[J]. Hydrogen Energy, 2008(33):1465–1470.

[4] Xing DF, Ren NQ, Wang AJ, Li QB, Feng YJ, Ma F. Continuous hydrogen production of auto-aggregative Ethanoligenens harbinense YUAN-3 under non-sterile condition[J]. Hydrogen Energy, 2008(33):1489–1495.

[5] Mu Y, Yu HQ, Wang G. Evaluation of three methods for enriching H2-producing cultures from anaerobic sludge[J]. Enzym Microb Technol, 2007(40):947–953.

[6] Wu SY, Hung CH, Lin CY, Lin PJ, Lee KS, Lin CN, et al. HRT-dependent hydrogen production and bacterial community structure of mixed anaerobic microflora in suspended, granular and immobilized sludge systems using glucose as the carbon substrate[J]. Hydrogen Energy, 2008(33):1542–1549.

[7] Zhang ZP, Show KY, Tay JH, Liang DT, Lee DJ. Biohydrogen production with anaerobic fluidized bed reactors-A comparison of biofilm-based and granule-based systems[J]. Hydrogen Energy, 2008(33):1559–1564.

[8] E.W.J. van Niel, M.A.W. Budde, G.G. de Haas et al. Distinctive properties of high hydrogen producing extreme thermophiles, Caldicellulosiruptor saccharolyticus and Thermotoga elfii [J]. International Journal of Hydrogen Energy, 2002(27): 1391-1398.

[9] Nandi R, Sengupta S. Microbial production of hydrogen: an overview [J]. Crit Rev Microbiol, 1998(24):61-84.

[10] Rachman MA, Nakashimada Y, Kakizono T, Nishio N. Hydrogen production with high yield and high evolution rate by self-Locculated cells of Enterobacter aerogenes in a packed-bed reactor[J]. Appl Microbiol Biotechnol, 1998(49):450-454.

[11] PEP Koskinen, CH Lay, JA Puhakka et al. High-efficiency hydrogen production by an anaerobic, thermophilic enrichment culture from an Icelandic hot spring[J]. Biotechnology and Bioengineering, 2008(101):665-678.

(330a) Adsorption of the Compounds Encountered in Monosaccharide Dehydration in Zeolites

Leon, M., University of Delaware
Xiong, R., University of Delaware
Swift, T. D., University of Delaware
Nikolakis, V., University of Delaware
Sandler, S. I., University of Delaware
Vlachos, D. G., University of Delaware
Originally presented on: 11/5/2013 12:30:00 - 12:52:00

Biomass carbohydrates are the most abundant renewable resources available and have the potential to replace petroleum as a source of both energy and chemicals, leading to more environmentally friendly processes and reducing the dependence on crude oil. A promising reaction in this field is the acid-catalyzed dehydration of monosaccharides, such as fructose and glucose, towards 5-hydroxymethylfurfural (HMF), an important intermediate for the production of valuable chemicals.[1] When zeolites are used as catalysts the preferential adsorption of reactants, products or byproducts could markedly affect the yield to the desired product. In this regard, studying the adsorption of the compounds present in the reaction media in the zeolite catalyst becomes crucial to understand the reactivity of the system and it is necessary to calculate the intrinsic reaction rate constants from kinetic data. Zeolites can also be used as adsorbents for the selective removal of the HMF from the reaction media (either in-situ or after dehydration). In this respect understanding the effect of the framework type, composition (e.g. Si/Al, extraframework cation), and presence of co-solvent (e.g. dimethylsulfoxide (DMSO)) on HMF adsorption are important for the design of an adsorptive separation process.

Adsorption loadings from solution are usually estimated from changes of the solute concentration assuming that the solution volume is not affected by adsorption. This quantity is called excess adsorption loading and sometimes it is quite different from the real loading. We formulated and used a method for the estimation of the real adsorption loading from the experimentally measured excess adsorption that can be applied to calculate the adsorption isotherms both in the case of single-solute and multi-solute mixtures.

The effect of temperature on the adsorption of fructose, glucose, HMF, levulinic and formic acid from aqueous solutions in the protonated form of zeolite beta with SiO_2/Al_2O_3 ratio of 18 (H-BEA) will initially be presented. The limiting heat of adsorption at zero coverage of all components was calculated from isotherms measured at 0, 25 and 40 °C. Adsorption experiments from multi-solute solutions were also carried out and the data were found to be in good agreement with predictions made using Ideal Adsorbed Solution Theory (IAST) and the single solute isotherm experimental data.

Adsorption of HMF was studied in more detail. We examined the effect of the framework type (BEA, MFI, FAU), composition (Si/Al) as well as the addition of DMSO as co-solvent on its adsorption uptake. It was found that in each zeolite type the adsorption uptake of HMF increases with the hydrophobicity (higher Si/Al ratio) of the zeolite. Experimentally measured isotherms were in good agreement with those calculated using Grand Canonical Monte Carlo - Expanded Ensemble (GCMC-EE) simulations. It is found that in the case of HMF/water binary mixtures, zeolites with larger pores (i.e. FAU), have higher HMF adsorption capacity but lower selectivity because large pores allow water adsorption due to favorable H_2O-HMF attraction. Therefore, dealuminated zeolite Y is less selective for the separation of HMF from water than silicalite-1 (siliceous MFI type zeolite) that has smaller pores sizes. In addition, experiments and simulations show that in HMF/DMSO/water ternary mixtures, the HMF loading decreases with increasing DMSO concentration. This decrease is due to the high solubility of HMF in DMSO, and to the fact that in the entropic contribution to the free energy of adsorption dominates over the enthalpic contribution. Analysis of our data shows that the ratio of the energetic interaction in the zeolite to the solvation free energy is a key factor in controlling separation from liquid mixtures.

To conclude, our findings could shed light on the acid catalyzed dehydration of monossacharides and the HMF separation, and serve as a hint to the rational design of microporous adsorbents for liquid-phase separations in biomass processing.

REFERENCES

1. Chheda, J. N.; Huber, G. W.; Dumesic, J. A. Liquid-Phase Catalytic Processing of Biomass-Derived Oxygenated Hydrocarbons to Fuels and Chemicals. *Angew. Chem. Int. Ed.* **2007,** *46*, 7164-7183.

2. Vandam, H. E.; Kieboom, A. P. G.; Vanbekkum, H. The Conversion of Fructose and Glucose in Acidic Media - Formation of Hydroxymethylfurfural. *Starch-Starke* **1986,** *38*, 95-101.

3. Kruger, J. S.; Nikolakis, V.; Vlachos, D. G. Carbohydrate Dehydration Using Porous Catalysts. *Curr. Opin. Chem. Eng.* **2012,** *1*, 312-320.

4. Kruger, J. S.; Choudhary, V.; Nikolakis, V.; Vlachos, D. G. Elucidating the Roles of Zeolite H-Bea in Aqueous-Phase Fructose Dehydration and HMF Rehydration. *ACS Catal.* **2013**.

5. Roman-Leshkov, Y.; Chheda, J. N.; Dumesic, J. A. Phase Modifiers Promote Efficient Production of Hydroxymethylfurfural from Fructose. *Science* **2006,** *312*, 1933-1937.

(330b) Elucidating Structure-Properties Relations for the Design of Highly Selective Carbon-Based HMF Sorbents

Yoo, W. C.,
Rajabbeigi, N., University of Minnesota
Mallon, E. E., University of Minnesota
Tsapatsis, M., University of Minnesota
Snyder, M. A., Lehigh University
Originally presented on: 11/5/2013 12:52:00 - 13:14:00

Continuously increasing energy demands in a fragile global market and in the face of growing environmental concerns motivate the drive to develop sustainable energy resources like plant biomass for replacing or augmenting current petroleum-derived fuels and chemicals [1-3]. One of numerous challenges facing the biorefinery is the selective and efficient production of important chemical products such as 5-hydroxymethyl furfural (HMF). HMF production from fructose dehydration in DMSO with various catalytic systems has shown superior selectivity and high yields [3, 4]. Yet, the subsequent need for energy-intensive separation of HMF from HMF/fructose/DMSO mixtures challenges the practical feasibility of this process. Microporous carbon materials have shown promise as substrates for alternative adsorption-based separations [5], but the origin of HMF selectivity and adsorption capacity, and thereby structure-properties relations enabling rational design of enhanced sorbents, has remained elusive. Through systematic quantification of the functionality and texture of various commercial (BP2000, Norit1240) and synthetic carbons (three-dimensionally ordered macroporous carbons; mesoporous carbons), we link, for the first time, adsorption capacity and HMF adsorption selectivity to two tunable materials properties of the carbon sorbents: micropore texture and oxygenate functionality (i.e., carbon polarity). In the process, we exploit these newly elucidated structure-properties relations for realizing a class of synthetic carbons with sub-micron particulate morphology that achieve more than 60% higher selectivity and more than 20% higher capacity for HMF over fructose as compared to the best performing commercial product, BP2000.

References

[1] P.Y. Dapsens, C. Mondelli, J. PÃ©rez-Ramírez, ACS Catal., 2 (2012) 1487-1499.

[2] G.W. Huber, S. Iborra, A. Corma, Chem. Rev., 106 (2006) 4044-4098.

[3] A. Corma, S. Iborra, A. Velty, Chem. Rev., 107 (2007) 2411-2502.

[4] X. Qi, M. Watanabe, T.M. Aida, R.L. Smith Jr, Ind. Eng. Chem. Res., 47 (2008) 9234-9239.

[5] N. Rajabbeigi, R. Ranjan, M. Tsapatsis, Microporous Mesoporous Mater., 158 (2012) 253-256.

(330c) Separation of Lignin From Hydrolyzates By Flocculation Using PEO Polymers

Ramarao, B. V., ESPRI, SUNY College of Environmental Science and Forestry
Yasarla, R., State University of New York
Gnanavel, V., SUNY ESF
Originally presented on: 11/5/2013 13:14:00 - 13:36:00

Hydrolyzates produced by pretreatment of lignocellulosic materials contain phenolic compounds, organic acids and aldehydes such as furfural in colloidal and dissolved form. Since these compounds can be potent inhibitors for fermentation, they need to be separated and isolated ahead of such biological treatments. We found that the application of a strongly binding non-ionic polymer, polyethylene oxide causes flocculation and rapid separation of lignin from the hydrolyzates, while preserving the carbohydrates in solution to a large extent. The effect of flocculation is measured by the reduction in turbidity and enhancement of the suspension settling velocity. The maximum flocculation was found to depend on the dosage of the polymer and its MW, but was independent of the pH of the hydrolyzates (from 2 through 9). PEO dosages were between 20 to 50 ppm at the maximum flocculation and reduction in suspension turbidity. The suspension temperature plays an important role, with temperatures below 21.5 C showing no flocculation.

The hemicellulose fraction in the supernatants was close to the raw extract showing that flocculation does not remove significant amount of fermentable sugars. Hence such pretreatment clarification is not likely to affect downstream bioproduct yields. Flocculation and clarification with PEO were also demonstrated on large pilot scale batches of 1000 kg of extracts. The flocculated extract was filtered and the filtrate showed a 99.5% reduction in turbidity, from 12000 NTUs in the raw extract to 50 NTUs in the filtrate. The impact of flocculation on solid liquid separations was quantified and flocculation was optimized with respect to dosage, suspension concentration, temperature and pH.

(330d) Fractionation and Recovery of Lignin and Xylan From Alkali Liquors - Characterization for Biorefinery Processes

Stoklosa, R. J., Michigan State University
Hodge, D., Michigan State University
Originally presented on: 11/5/2013 13:36:00 - 13:58:00

Understanding the behavior of plant cell wall biopolymers during alkaline biorefining processes and how their properties are affected by the processing conditions is important for maximizing value from woody lignocellulosic feedstocks. In this work, the fractionation of alkali pulping liquors was investigated to characterize the extractability, recoverability, and properties of the solubilized hemicelluloses and lignin during the alkali impregnation and heat-up stages of alkali organosolv pulping or soda pulping for three hardwood species: sugar maple (*Acer saccharum*), silver birch (*Betula pendula*), and a hybrid poplar (*Populus nigra x maximowiczii cv.* NM6). Recoverability of the solubilized biopolymers were determined using a variety of methods including aggregation and precipitation by acidification as well as ultrafiltration. The xylan showed a decreasing degree of polymerization (DP) with increasing temperature during reactor heat-up, which stabilized at a DP between 10-30. Lignin aliphatic and aromatic hydroxyl content was characterized by ^{1}H NMR and the lignin molecular weight distributions were determined by gel permeation chromatography (GPC). These properties showed a clear dependence on the extraction conditions and were clearly linked to recoverability of lignin by either acidification or ultrafiltration.

(330e) Novel Catalytic Membranes for Combined Biomass Hydrolysis and Sugar Recovery

Avram, A., University of Arkansas
Lei, J., University of Arkansas
Qian, X., University of Arkansas
Wickramasinghe, S. R., University of Arkansas
Beier, A., University of Arkansas
Originally presented on: 11/5/2013 13:58:00 - 14:20:00

Production of biofuels from lignocellulosic biomass is one of the leading candidates for replacement of fossil fuels. In the biochemical conversion platform, dilute sulfuric acid pretreatment is one of the leading technologies to hydrolyze hemicellulose, relocate lignin and expose cellulose. After pretreatment, a cocktail of cellulase enzymes is used to break down cellulose synergistically. The major problems associated with this technology are the slow and incomplete hydrolysis of cellulose and the degradation of monomer sugars during pretreatment. Furthermore, the cost of the enzymes has been an inhibitory factor for the commercialization of biomass conversion technology.

Here, a novel catalytic membrane reactor is designed to simultaneously hydrolyze cellulose and hemicellulose and separate the hydrolyzed sugars, thus minimizing product degradation. A polymeric solid acid catalyst is synthesized to replace the corrosive sulfuric acid and expensive enzymes, but operates at a higher conversion rate than the analogous natural enzymes and that costs far less to produce. We have grown adjacent nanostructure consisting of tethered polysulfonic acid and polyionic liquid chains from the surface of ceramic membranes. The feed consisting of cellulose and hemicellulose in a water / ionic liquid mixture is introduced on one side of the membrane. The polyionic liquid chains help solubilize the polysaccharides while the polysulfonic acid chains catalyze the hydrolysis of the polysaccharides. The membrane pore size is such that only water containing dissolved monomeric sugars and short chain oligomers can pass through. The ionic liquid and larger polysaccharides are rejected.

From preliminary hydrolysis experiments, our results indicate the potential of these novel membranes to hydrolyze polysaccharides, minimize degradation product and save on catalyst costs.

(330f) Optimization of An Extractive Distillation With Mixed Solvents As Entrainer

Garcia, L. C., Universidad Nacional de Colombia
Gil, I. D., Universidad Nacional de Colombia
Orjuela, A., Universidad Nacional de Colombia
Rodriguez, G., Universidad Nacional de Colombia
Originally presented on: 11/5/2013 14:20:00 - 14:42:00

Anhydrous ethanol is widely used in the biofuels industry, since it has demonstrated to be environmental friendly and to have excellent properties for clean-burning. In mixtures of ethanol and gasoline, alcohol must have purity higher than 99.5% in order to avoid phase separation. Among the most popular processes used in ethanol dehydration, extractive distillation has demonstrated to be an adequate alternative due to the low energy consumption compared with other techniques and its economic feasibility. In this work, an extractive distillation process was simulated with the Aspen Plus software platform, using the RadFrac module for distillation columns. A mixture of ethylene glycol and glycerol is proposed as entrainer. This enhanced entrainer combines the low viscosity and the ability of the ethylene glycol to break the binary azeotrope with the low prices of glycerol. Phase equilibrium of this quaternary system was calculated with the NRTL model, which correlates well these strongly polar mixtures. The nominal case consists in the dehydration column, the recovery column and the solvent recycle loop. In order to establish the most adequate operating conditions, it was studied for each column the effect of the molar reflux ratio, the feed stage and the feed condition in the ethanol and solvent compositions. For the dehydration column, it was also investigated the effects of the ethylene glycol-glycerol mixture composition, the separating agent feed stages, the separating agent split stream, and the azeotropic feed temperature on the separation. Subsequently, the thermal and hydraulic analysis of both columns was established, so that the system operates energetically efficient. Finally, the economic optimization of the system was achieved, using the maximization of the profit as objective function. In this case, the solvent to feed ratio, the reflux ratio and the number of stages in the extractive column were the optimization variables, while the design parameters of the recovery column were fixed.

(330g) Study of Physical Properties of Binary Mixtures of Sulfolane + 2-Butanol (or + 2-propanol) At Different Temperatures

Mesquita, F., Federal University of Ceara
Sant'ana, H. B., Universidade Federal do Ceara
Feitosa, F., Universidade Federal do Ceara
Aznar, M.,
Santiago-Aguiar, R. S., Universidade Federal do Ceara
Originally presented on: 11/5/2013 14:42:00 - 15:04:00

Sulfolane and alcohols are classes of chemical compounds that are usually used as a solvent in different industrial recovery processes. The focus of the present work is to measure volumetric (densities) and transport (dynamic viscosity) properties of the following binary mixtures: sulfolane + 2-butanol (or + 2-propanol) at a range temperature from T = 303.15 to T = 343.15 K, at atmospheric pressure. Densities and viscosities were measured using a viscodensimeter (Anton Paar SVM 3000 Digital Oscillation U-tube). Viscosity deviations and excess volumes were obtained from the experimental data, by using a fitting procedure (Redlich-Kister type equation). The results show that excess volumes were positive over the entire mass fraction range for all binaries mixtures studied. Otherwise deviation viscosities are negative over the entire mass fraction range of the mixture. Excess volumes and deviation viscosities behaviors increasing with temperature.

(338a) Hierarchically Structured Nanomaterials for High Density Gas Storage and Compression

Schwartz, N. R., Mainstream Engineering Corporation
Chester, G. E., Mainstream Engineering Corporation
Hill, J. J., Mainstream Engineering Corporation
Originally presented on: 11/5/2013 15:15:00 - 15:35:00

The need for safe, transportable, and dense gas storage is growing, especially in the hydrogen and oxygen markets. A fabrication method for hierarchically structured nanomaterials and their use as gas storage vessels is currently being developed at Mainstream Engineering. The nanomaterials are formed by self-assembly and sealed to maintain high pressure with a material that also acts as controllable and reversible pump. Based on hydrostatic loading we have demonstrated that hydrogen can be stored at densities exceeding 0.12 kg H_2/kg material and 0.60 kg H_2/L material. Similarly, oxygen can be stored at 0.60 kg/kg and 1.08 kg/L. We predict that the volumetric and gravimetric storage density can exceed DoE's 2015 targets for hydrogen by 65% and 98%, respectively and with a safety factor of 4. These materials exceed the storage capacity of current technologies for other gasses as well. We will discuss the current development effort as well as the technological road map to prototype demonstration. In addition we will discuss the fundamentals of the gas storage, sealing and reversible pumping mechanism.

(338b) Prebridge Technique to Incorporate Transition Metal to Metal-Organic Frameworks for the Purposes of Catalysis and Hydrogen Storage Via Spillover Effect

Wang, C. Y., the Pennsylvania State University
Gong, Q., Rutgers University
Li, J., Rutgers University
Lueking, A. D., The Penn State University
Originally presented on: 11/5/2013 15:35:00 - 15:55:00

Metal-organic frameworks, high surface area materials with the characteristics of adjustable structure and surface chemistry, are becoming promising candidates for hydrogen storage, gas separation, and catalyst supports. In this paper, we explore MOF stability after various catalytic doping methods for IRMOF-8, Cu-BTC, and Cu-TDPAT. After identifying a catalytic doping technique that maximizes stability and retains the surface area of the MOF precursor, we measure hydrogen isotherms at 300K up to 80 bar, and demonstrate catalyst addition significantly increases hydrogen storage via the hydrogen spillover effect. Adsorption enhancement is most pronounced at low pressure, and kinetic limitations and MOF instability effects inhibit high-pressure adsorption via the spillover effect. Novel hydrogen chemisorption sites are identified using spectroscopic techniques, for both undoped and doped MOFs. An improved mechanistic understanding of the hydrogen spillover effect is developed by tracking of hydrogenation of N groups via spectroscopy, density functional theory, and comparison of defected versus pristine MOF structures. Implications for hydrogen storage and use of MOFs as a catalyst support are discussed.

(338c) Cryogenic Hydrogen Adsorption On MOF-5 With Novel Microchannel Thermal Management System - Experiment and Simulation

Loeb, C., Oregon State University
Jovanovic, G. N., Oregon State University
Originally presented on: 11/5/2013 15:55:00 - 16:15:00

Achieving a high volumetric storage density of hydrogen gas is a major challenge in the progress towards hydrogen-fueled automobiles. Of the various storage methods, physical adsorbents show great potential, however, a reliable thermal management system is vital to provide quick, safe, and efficient charging and a controlled release of hydrogen gas on demand.

Experimental investigations have focused on the adsorption of hydrogen gas in a compressed adsorbent bed under cryogenic conditions. The temperature distribution within the storage media is monitored and the storage capacity is determine to evaluate the affect of the addition of microscale-based heat exchanger device within the adsorption environment. A mathematical model is used to numerically simulate the experimental process and evaluate system performance under various initial and operating conditions.

A novel microscale-based heat exchanger design takes advantage of 0.250 mm high features, which facilitate the flow of cooling fluid. The heat exchanger is fabricated by diffusion bonding of patterned stainless steel shims, resulting in an overall 0.250 mm thick resistance to thermal transport. The microchannel structure uniformly distributes hydrogen gas into the adsorbent bed ensuring uniform adsorption and heat release.

Liquid nitrogen flow through the microchannel device improves the storage capacity of a MOF-5 adsorbent system by 5% compared to non-cooled experiments, while providing a significant gain in storage capacity in comparison to a cryo-compressed storage system.

An expanded system investigation containing multiple adsorbent beds in contact with a 'numbered-up' microchannel heat exchanger system is tested under cryogenic conditions to ensure uniform flow and temperature distribution through each plate during the adsorption process.

(338d) Core-Shell Ferrite Nanoparticles for Hydrogen Production From Thermochemical Water-Splitting Process

Amar, V. S., South Dakota School of Mines and Technology
Puszynski, J. A., South Dakota School of Mines and Technology
Shende, R., South Dakota School of Mines and Technology
Originally presented on: 11/5/2013 16:15:00 - 16:35:00

Hydrogen can be efficiently produced by the two-step thermochemical water splitting process which involves partial reduction of redox materials (e.g. ferrites) at elevated temperatures creating oxygen vacancies followed by oxidation of partially reduced redox material at lower temperatures leading to oxygen scavenging from steam and producing hydrogen. We report the sol-gel synthesis of core-shell nanoparticles of different ferrite materials which include nickel, manganese, zinc and cobalt ferrites. As prepared gel was aged for 48 hours and dried at about 100 ^{o}C for 2 hours and calcined at 800 ^{o}C. These core-shell nano particles were analyzed by Transmission Electron Microscopy (TEM), Brunauer-Emmett-Teller (BET), X-Ray Diffraction (XRD) to understand core-shell morphology, specific surface area and phase formation, respectively. The core-shell nanoparticles were loaded inside a tubular reactor as a packed bed with supporting material and heated to 700 to 1100 ^{o}C under the constant nitrogen flow rate of 35 to 100 sccm. After the activation step, water was introduced at 500 ^{o}C and the hydrogen in the product gas stream was continuously monitored using the online hydrogen sensor interfaced with a computer. Multiple thermochemical cycles were performed and the hydrogen volume generated for the ferrite core-shell nanoparticles was compared with ferrite nanoparticles. After the thermochemical cycles, the powdered material was recovered and analyzed by SEM/TEM, XRD, and BET. The results on the hydrogen volume generation obtained using ferrite core-shell nanoparticles will be presented.

(338e) Fabrication of Photovoltaic Thermal Water Electrolyzer for Hydrogen Production

Oruc, M. E., University of Illinois at Urbana-Champaign
Nuzzo, R. G., University of Illinois at Urbana-Champaign
Originally presented on: 11/5/2013 16:35:00 - 16:55:00

Solar energy is often harnessed by using a photovoltaic/thermal (PVT) collector, where a photovoltaic (PV) module and a thermal collector convert solar energy into electricity and heat, respectively. The heat energy has been conventionally used for hot water and/or room heating; it is, however, highly desirable to convert the released energy into alternative storable forms. Hydrogen is environmentally friendly and a sustainable energy carrier that can be stored. Currently the PV-assisted water electrolyzers available on the market have PV conversion efficiencies of only 15%. The energy that is not transformed into electricity is just dissipated. To this end, we have developed a new integrated system – so-called "photovoltaic thermal water electrolyzer" – which consists of PV cells positioned on top of a planar micro water electrolyzer. Herein, we provide the design, fabrication and characterization of a planar micro water electrolyzer for hydrogen production. The heat that is released from the system is used to raise the temperature of the electrolyte solution, thereby increasing the efficiency of the hydrogen production.

Analysis of the proposed design is performed using the COMSOL Multiphysics software package. Platinum black electrodes are deposited on a glass substrate and polydimethylsiloxane (PDMS) is used as the material for the electrolyzer chamber. An insulated flexible heater is attached to the top of the device and a constant temperature is applied in order to achieve the desired temperature of the electrolyte solution. Electrochemical characterizations are performed in 1 M H_2SO_4 solution at up to 70°C under applied voltages of 1.8 V and 2 V. Enhanced gas separation and management are also investigated in the chamber in order to increase hydrogen evolution. This approach produces hydrogen and oxygen gases at least four times faster relative to room temperature operation. In summary, this new integrated approach brings advantages over current systems such as utilization of wasted heat into efficient hydrogen production, a smaller spatial footprint with architectural uniformity that is scalable to all types of PV modules, and can lead to reduced installation costs.

(338f) High Temperature Steam Electrolysis in Hydrogen Production

Chen, C.,

Originally presented on: 11/5/2013 16:55:00 - 17:15:00

With the depletion of fossil fuel resources and the need to reduce greenhouse gases emission, hydrogen becomes a potential energy carrier in the future. There are three methods widely used in commercial scaled hydrogen production in the US. A comparison of the advantages and shortages of these methods has been listed. Among these technologies, high temperature steam electrolysis (HTSE) presents impressive performance on multiple aspects, such as operation efficiency, economic values and environmental impacts. Several HTSE applications in hybrid energy system have been presented. Potential improvements and recommendations on HTSE performance have also been proposed.

(338g) Hydrogen Production By Methanol Reforming in the Gradient Micro-DBD Plasma Reactor

Baowei, W. Sr., tianjin University
Wenjie Ge, G. Sr.,
Xiaofei, D.,
Originally presented on: 11/5/2013 17:15:00 - 17:35:00

Hydrogen production by methanol reforming in the gradient micro-DBD plasma reactor

Baowei Wang*,Wenjie Ge, Xiaofei Duan

(Key Laboratory for Green Chemical Technology of the Ministry of Education, School of Chemical Engineering and Technology, Tianjin University, Tianjin, 300072, China)

Abstract

The rise in global energy demands and the shortage of fossil fuels as well as environmental considerations have created a pressing need to seek renewable alternatives. Solar, wind, and biomass are all promising renewable resources but they are generally site-specific, intermittent and unstable. On the other hand, hydrogen has been identified as an ideal energy carrier to support sustainable energy supply and reactant due to its clean and efficient. However, to support a feasible hydrogen-based economy, it is essential to produce hydrogen cleanly and renewably [1].

Hydrogen can be manufactured from a variety of energy sources. For example, it can be produced from hydrogen carriers, such as methanol and ethanol. Methanol has merits of being liquid phase, having high hydrogen to carbon ratio and getting from biomass.

In recent years, many types of non-thermal plasma reactors for hydrogen production have been implemented such as gliding arc micro-reactor [2], dielectric barrier discharge micro-reactor [3] and microhollow cathode discharge reactor [4]. These reactors processes may provide significant advantages in terms of fast reactions and energy efficiency. But the selectivity for the goal products and the conversion of the methanol are not considerable high.

A new design of a gradient micro-DBD (Gm-DBD) plasma reactor in figure 1 (2#, 3#) is preliminarily investigated for the reforming of methanol to various useful products. In methanol reforming by Gm-DBD reactor under ambient conditions, hydrogen and carbon monoxide are dominantly produced. The conversion of methanol reaches 81.3%, which is much higher than this in the conventional micro-DBD reactor in figure

1 (1#). The changes of electron temperature and electron density distribution, the high energy density can get in Gm-DBD reactor, as compared to conventional micro-DBD reactor.

The results show that the conversion of methanol depends on the form of gradient electrode. Our previous woke proved that the methanol reforming in the plasma reactor is depend on energy density and residence time [5]. The electrons generated in high energy density have more energy to break down the strong C-O bond in methanol, leading to more decomposition of methanol molecules. The long residence time make more contact time for the reactant species to collide with electrons and more time to the free radicals recombination in the reactor. In the Gm-DBD reactor the discharge gap effects of energy density and residence time. The energy density increases with the decreases of discharge gap and the residence time increases with the discharge gap. The highest conversion of methanol is found in 3# reactor. Firstly, the discharge gap is small, the energy density is high, more methanol molecules decompose to generate of more desirable species. Then the discharge gap increase, since discharge space and residence time increase with electrode gap, the number of collisions among the methanol molecules and the electrons, at the same time, the free radicals can recombine in the reactor, completely.

From the results, we found that the high energy density make more methanol molecules decompose. The long residence time lead more contact time for the reactant species to collide with electrons and more time to the free radicals recombination in the reactor. Under this operating condition, the highest methanol conversion can be got.

References

[1] Profeti L P R, Dias J A C, Assaf J M, et al. Hydrogen production by steam reforming of ethanol over Ni-based catalysts promoted with noble metals. J Power Sources, 2009, 190(2): 525-533.

[2] Rueangjitt N, Sreethawong T, Chavadej S, et al. Plasma-catalytic reforming of methane in AC microsized gliding arc discharge: Effects of input power, reactor thickness, and catalyst existence. Chem Eng J, 2009, 155(3): 874-880.

[3] Nozaki T, Ağıral A, Yuzawa S, et al. A single step methane conversion into synthetic fuels using microplasma reactor. Chem Eng J, 2011, 166(1): 288-293.

[4] Lindner P J, Besser R S. A microplasma reactor for chemical process intensification [J]. Chem Eng & Tech, 2012, 35(7): 1249-1256.

[5] LÜ Y, Yan W, Hu S, et al. Hydrogen production by methanol decomposition using gliding arc gas discharge[J]. J of Fuel Chem & Tech, 2012, 40(6): 698-706.

* To whom Correspondence should be addressed. E-mail□wangbw@tju.edu.cn ,Tel□86-22-27402944, Fax□86-22-87401818.

(338h) H2 Storage in Adsorbents With Various Porous Structure
Wang, K.,
Originally presented on: 11/5/2013 17:35:00 - 17:55:00

Hydrogen storage is important for the 'hydrogen economy' and the sustainable development for the world [1]. Hydrogen storage was investigated in six nanoporous adsorbents, i.e. two template-synthesized carbons, a super-high surface area carbon (AX-21), a zeolite (NaY), and the ETS-10.

The surface textures of these adsorbents, including area and micropore volumes, were fully characterized [2] and compared while their hydrogen storage capacities were measured at low pressure (1bar), mid-pressure (20bar), and high pressure (80bar), respectively. It was found that the template-synthesized carbon TC-1 and the commercial super-high surface area carbon AX-21 present the best performance at both the low and the high pressures (Figure 1). The relationship between the adsorption capacity, micropore volume as well as the surface areas were discussed [3] . The density and micropore volume were found to be the most critical factors for the high H_2 storage capacity, while the surface area plays the next important role [4].

(420a) Drop-in Fuels Production From Biofeedstocks Using the Current Refinery Infrastructures

Feng, M., Southwest Research Institute

Originally presented on: 11/6/2013 8:30:00 - 8:52:00

This paper will give an overview of the current status of drop-in fuels production from biofeedstocks, including vegetable oils, plant seed oils, biomass pyrolysis oils, and microalgae oils. Blending the biofeedstocks into the hydrocarbon streams in the current refinery infrastructures provides great advantages for future refining industries, including: 1) minimizing CAPEX cost; 2) taking the advantage of synergy effect between refining of conventional oxygen-deficient hydrocarbon feedstocks and oxygen enriched biofuels; 3) reducing emissions and CO_2 footprint. Biofeedstocks stabilization is necessary. Revamping of the current refinery infrastructure for such co-processing will increase flexibility of the refining industry. This paper will also address the pretreatment requirements for biofeedstocks and guard bed design for the hydrotreating unit. A lot of challenges still remain in catalyst design and short life-time for hydrotreating of oxygenated compounds by hydrodeoxygenation and hydrodecarboxylation. Handling different heteroatoms, especially phosphorus contaminants in the feedstock originated from phospholipids in biomass feedstock, is still in learning curve to the refinery.

(420b) Saccharification of Miscanthus x Giganteus With Dilute Nitric Acid

Yang, F., UC Berkeley
Afzal, W., UC Berkeley
Liu, N., UC Berkeley
Cheng, K., UC Berkeley
Pauly, M., UC Berkeley
Bell, A. T., University of California, Berkeley
Prausnitz, J. M., UC Berkeley
Originally presented on: 11/6/2013 8:52:00 - 9:14:00

Miscanthus × giganteus (M. giganteus), a C_4 perennial grass, is a potential renewable resource for energy and chemicals. *M. giganteus* contains 41.5% cellulose, 24.5% hemicellulose, 27.0% lignin, 7% others (4% ash and 3% extractables). Hydrolysis of hemicellulose and cellulose gives sugars that can be converted to a biofuel or to useful chemicals. Like similar agricultural materials, *M. giganteus* contains lignin that makes hydrolysis difficult.

We investigated a two-step dilute nitric acid process at advanced temperatures to obtain sugars from *M. giganteus*. The two-step process removes hemicellulose using dilute nitric acid at 120-160 °C for 10 to 40 min whereas cellulose hydrolysis is conducted at higher temperature, 170-200 °C for less than 6 min. Nitric acid is a better choice than sulfuric acid due to its compatibility with stainless steel. It can easily be neutralized with ammonia to produce ammonium nitrate, a useful nutrient for microorganisms used in downstream fermentation. We present the best conditions for dissolving about 75% polysaccharides (cellulose and hemicellulose together) keeping degradation minimal. We also present chemical analysis of the solid and liquid phase. In the first step, more than 90% of hemicellulose was transferred to the liquid phase and overall for both steps, more than 70% of cellulose was transferred to the liquid phase.

For better understanding of the process, we investigated the effects of various conditions by varying the temperatures, acid concentration, solid-to-liquid ratio and residence time. High-performance liquid chromatography was used to determine any degradation of sugars in the liquid phase; we observed small amounts of furfural and 5-(hydroxymethyl)furfural. In addition, the recovered solid *M. giganteus* was characterized by Fourier-transform infrared spectroscopy and by two-dimensional ^{13}C-^{1}H heteronuclear single quantum coherence nuclear magnetic resonance spectroscopy, scanning-electron microscopy, attenuated total reflection-infrared spectroscopy and X-ray diffraction. 2D-NMR analysis shows that a higher temperature enhances the removal of hemicellulose. Dilute nitric acid pretreatment reduced 9% lignin in *M. giganteus* at 140 °C and 20 min while it decreased 27% lignin at 160 °C and 10 min. A substantial amount of lignin side-chains, mostly β-O-4' aryl ether and 5-5'/α-O-1' dibenzodioxin, were cleaved. As a result, the molecular weight of lignin in recovered residues is lower than that in native lignin.

(420c) Pyrolysis Kinetics of Anisole a Simple Lignin Model Compound

Koirala, Y.,

Villano, S., Colorado School of Mines

Carstensen, H. H., Laboratory for Chemical Technology

Dean, A. M., Colorado School of Mines

Originally presented on: 11/6/2013 9:14:00 - 9:36:00

Pyrolysis Kinetics of Anisole a Simple Lignin Model Compound

Abstract

The thermal decomposition of lignin, one of the major components of biomass, is not well characterized and leads to formation of undesirable products that require substantial upgrading before using it as a drop-in fuel. An improved understanding of the underlying decomposition reactions of lignin could provide insight into process control conditions that could mitigate the formation of these undesirable products during the fast pyrolysis process to produce biofuels. Because the structure of lignin is so complex, model compounds are often utilized to gain insight into this underlying chemistry. The simplest model that can be employed to investigate the weak phenolic-carbon linkage in lignin is anisole ($C_6H_5OCH_3$). Although several studies have been carried out to understand the decomposition of anisole, there are limited data available under fast pyrolysis process conditions. Furthermore, the existing models have not been able to predict the observed product formation. In this work, we present anisole thermal decomposition data that was collected under conditions that closely resemble fast pyrolysis (T=525-650°C, t ~0.9s, and P~0.8 atm) using a tubular flow reactor. These data were modeled using a mechanism that was developed by Pecullan et al.[1] , but it was modified to include updated rate constants for the reactions involving the o-methylcyclohexadienone potential energy surface that were derived from electronic structure calculations.

The stationary points and transition states located on this PES were calculated using the CBS-QB3 level of theory using Gaussian 03.[2] To improve the accuracy of the D_fH_{298}, S_{298}, and C_p values, low frequency vibrational modes that resemble torsions around single bonds are treated as hindered internal rotors; the hindrance potentials were calculated at the B3LYP/6-31G(d) level of theory via relaxed surface scans. The electronic energy of each species was converted to its heat of formation using the atomization method. High-pressure rate coefficients were calculated using canonical transition state theory. These high pressure rate constants were used to obtain apparent pressure and temperature dependent rate constants using Quantum-Rice-Ramsperger-Kassel theory with a modified strong collision approximation.[3] This mechanism was used to model the data collected here as well as previously published data; the simulations were performed using the ChemKin-Pro software package.[4]

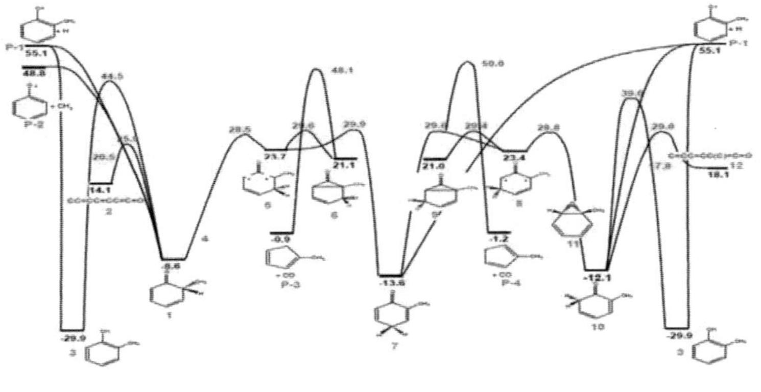

Figure 1. Potential Energy Surface for o-methyl cyclohexadienone (units: kcal mol^{-1})

A simplified version of o-Me*cy*HDOE potential energy surface calculated at the CBS-QB3 level of theory is shown in Figure 1. The barrier for isomerization to form o-cresol (3) is predicted to be ~53 kcal mol^{-1}, while those to form the two other methyl cyclohexadienone isomers (7 and 10) are much lower (~41 and ~43 kcal mol^{-1}). Note that these latter isomerization reactions proceeds through a diradical intermediate, which can eliminate CO to form 1- and 2-methyl cyclopentadiene (P3 and P4). The overall barriers for the formation of these two species are ~57 and ~63 kcal mol^{-1}, respectively. The barrier to form 1-methyl cyclopentadiene is comparable to that for ortho-cresol formation, suggesting that these channels might be of comparable importance. The surface also contains several reaction pathways to methyl phenoxy plus H products (P1). Although these reactions are ~6 kcal mol^{-1} higher in energy than the entrance channel (P2), these dissociation reactions are entropically favored when compared to isomerization. These channels are potentially important since these H atoms can add to methyl phenoxy radical to directly form o-cresol or they can add to phenoxy radical to form phenol.

The potential energy surface determined in this work has some important differences compared to the one originally estimated by Pecullan *et al.*[1] (Note that at the time that the Pecullan *et al.* mechanism was assembled, high level electronic structure calculations were generally not feasible for systems of this size. Thus, this PES and the corresponding high pressure rate constants were estimated in analogy to other similar reactions.) The CBS-QB3 barrier for ortho-cresol formation is ~14 kcal mol^{-1} higher in energy, while those for methylcyclopentadiene formation are at least 25 kcal mol^{-1} higher in energy. We find that o-methyl cyclohexadienone reacts to form the 1- and 2-methyl cyclopentadiene isomers, while Pecullan's analysis predicts the formation of the 5-methyl cyclopentadiene isomer.

Incorporation of these rate constants leads to improved predictions for many of the observed products. Selected comparisons are shown in Fig. 2. Although this updated mechanism is an improvement, it is clear that further progress is needed. Current efforts are focused on extending the type of analysis described here to other reactions that occur during anisole pyrolysis.

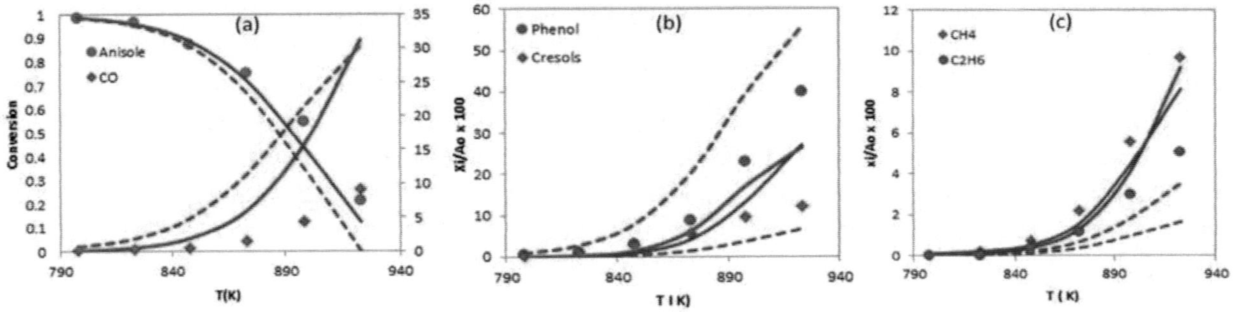

Figure 2. Comparisons of anisole pyrolysis products distribution data with model predictions where solid lines indicate the updated model and the dashed lines indicate the Pecullan model.

References

1. Pecullan, M,; Brezinsky, K,; and Glassman, I, *J. Phys Chem A.*, **1997**, *101*, 3305 ? 3316.
2. Frisch, M. J.; Trucks, G. W.; Schlegel, H. B.; Scuseria, G. E.; Robb, M. A. et al. Gaussian 03, Revision A. 1, Gaussian, Inc: Pittsburgh, PA, 2003.
3. Chang, A. Y.; Bozzelli, J. W.; Dean, A. M., *Z. Phys. Chem.* **2000**, *214*, 1533-68
4. ChemKin-Pro, Reaction Design, San Diego, 2008

(420d) A Catalytic Hydrotreatment Model for Plant Oils and Bio-Oils to Green Diesel

Mullins, M., Michigan Technological University
Shonnard, D. R., Michigan Technological University
Funkenbush, L.,

Originally presented on: 11/6/2013 9:36:00 - 9:58:00

Plant oils are triglycerides which are complex oxygenated mixtures of fatty acids. They may be reacted with alcohols in the presence of a base catalyst to yield fatty acid methyl ester (FAME) fuels; however, FAME fuels can only be blended in limited amounts with pure hydrocarbon fuels, may cause coking and have poor cold flow properties. Direct catalytic hydrotreatment (HDT) of plant oils to paraffinic hydrocarbons yields hydrocarbon components with a higher heating value, which are readily blended with traditional petroleum fuels in a product commonly referred to as green diesel. Catalytic hydrotreatment of plant oils is an exothermic process that can follow either a hydro-deoxygenation or decarboxylation route to lower the oxygen content, but both processes require substantial hydrogen. A product distribution similar to diesel fuel is usually desired, but the actual distribution is a strong function of the feed composition, reactor process conditions, and the type of catalysts employed. Unfortunately, the catalytic routes by which plant oils react with hydrogen and the reaction intermediates are not well understood. In our study, a multi-phase reactor model incorporating gas/liquid ratios, mass transfer, pressure, temperature, reactor residence time, and intrinsic reaction kinetics has been developed. This model, combined with the catalyst-specific kinetics, yields the stoichiometry, energy requirements and product distribution for a given feed material. The effects of temperature, pressure, hydrogen/oil ratio and reactor space velocity have received particular attention. As reaction temperature increases the yield of n-paraffins decreases, while the ratio of iso-paraffins increases. Lower operating pressures generally favor green diesel production, since higher pressures also promote cracking reactions. Similarly, conversion efficiency decreases with increasing hydrogen/oil ratio and cracking reactions are triggered, lowering paraffin yields. Higher reactor space velocities are preferable since they also promote $C_8 - C_{30}$ production. This model allows us to assess the process efficiency and economics and perform a realistic LCA for different plant oil feedstocks

(420e) Catalytic Deoxygenation of Butyric Acid Derived From Biomass for Production of Jet Fuel

Nahreen, S., Auburn University
Gupta, R. B., Auburn University
Originally presented on: 11/6/2013 9:58:00 - 10:20:00

Butyric acid can be produced by fermentation of biomass using Clostridium butyricum as biocatalyst and it is also produced as a by-product with acetone-n-butanol-ethanol (ABE) mixture in the well known 'ABE fermentation process'. Applying advanced fermentation method with integrated separation processes, a product concentration of 20 gl^{-1} was achieved and it can be further increased by using synthetic microbial strain which can overcome the product inhibition effect experienced in bio-butanol production. Butyric acid derived from biomass can be converted to drop-in hydrocarbon fuel by catalytic deoxygenation process using commercial acid catalysts as γ-Al_2O_3 and ZSM-5, Zeolite. At operating conditions of 400°C and 3 bars, conversion of pure butyric acid in a packed bed tubular reactor on γ-Al_2O_3 catalyst produces an organic liquid product phase immiscible with water at a very high yield of 80.5 wt%. The higher heating value (HHV) of the liquid organic product is 23.5 kJ/g. The organic liquid produced contains mostly saturated and unsaturated ketones which can be again deoxygenated to produce higher hydrocarbon molecules. With increasing temperature and decreasing weight hourly space velocity (WHSV) of the process, product composition and components can be altered and heavier hydrocarbon molecules can be produced in a single step. A two- step catalytic deoxygenation approach with series packed bed reactors having γ-Al_2O_3 and ZSM-5, Zeolite as catalysts respectively is compared with a one step catalytic deoxygenation process in terms of energy content of the final products and also total energy consumption of the entire process.

(420f) A Kinetic Model for the Enzymatic Hydrolysis of Lignocellulose for Ethanol Production

Rosales, O. Sr.,
McKenzie-Trajano, H. L.,
Duff, S., University of British Columbia
Posarac, D.,
Originally presented on: 11/6/2013 10:20:00 - 10:42:00

In order to make the lignocellulosic ethanol a commercially attractive industry and compete with the corn ethanol and oil industry, several points in the lignocellulose process needs to be improving. Among these challenge, the improvement of the enzymatic hydrolysis is one of the most important due to the high enzyme cost. The optimization of the enzymatic hydrolysis depends on the development of hydrolytic kinetic models that may accurately quantify the cellulose conversion, hydrolysis rate and the variables that slowdown the process. Furthermore, a robust model is needed to develop rate expressions that can be incorporated into process models which can be used to the size, optimize and techno-economic evaluate the lignocellulose ethanol production. Therefore, a kinetic model was developed for the enzymatic hydrolysis of lignocellulose that provides guidance on the adjustments of enzyme dosage and hydrolytic time. The model successfully predicted the production of glucose at different enzyme loadings (20 and 40 FPU/ g cellulose) and pretreatment conditions. The proposed model considers the lost of enzyme due to the lignin-enzyme interactions, which offers the opportunity to relate the pretreatment and hydrolysis processes. The developed model offers the opportunity to study the pretreatment and hydrolysis relation and to techno-economically evaluate the viability of different pretreatment and hydrolysis conditions for the ethanol production.

(420g) Biodiesel Production Using Continuous Flow Ultrasonic Reactor

Kumar, A., Indian Institute of Petroleum

Naik, D. V.,

Atrey, N., Indian Institute of Petroleum

Kaul, S.,

Originally presented on: 11/6/2013 10:42:00 - 11:04:00

Biodiesel production using continuous flow ultrasonic reactor

Ajay Kumar[a]*, D.V.Naik[a], Neeraj Atrey[a], S. Kaul[a]

[a]Indian Institute of Petroleum, Dehradun, India 248005.

*Corresponding author. Tel.:+911352525949 Fax: +911352660202

E-mail address: ajaykr@iip.res.in

Abstract

The soaring prices of petroleum, depletion in the petroleum reserves and the growing environmental concerns to limit the exhaust emissions have led to the emergence of biodiesel as an alternative to diesel fuel. Biodiesel has attracted enormous attention because it is a biodegradable, renewable source, sustainable, non-toxic and safe alternative for meeting the world?s energy needs that also reduces emission of carbon monoxide, carbondioxide, sulfur oxides, heavy metals, aromatic hydrocarbons and volatile compounds.

Biodiesel is traditionally obtained from vegetable oils or animal fats subjected to a chemical reaction called transesterification. In this reaction, the vegetable oil reacts in the presence of a catalyst and alcohol to produce the corresponding alkyl ester of the fatty acid mixture. Glycerol is produced as a by-product. Methanol and ethanol are the most commonly used alcohols because of their low cost. Both homogeneous and heterogeneous catalysts are used for biodiesel production.

It is understood that the transesterification of triglyceride with methanol is equilibrium reaction. The ratio of alcohol to triglyceride needs to be higher than stoichiometric to drive the equilibrium to a higher biodiesel yield. The transesterification reaction involves two immiscible phases. The less dense phase has the catalyst dissolve in the alcohol, where as the other contains the oil or fat. The reaction between these species can occur only in the interfacial region between the liquids, as alkaline catalysts are insoluble in the oil phase. Vigrous mixing is required to increase the area of contact between the two phases. Use of ultrasound reactor favors the reaction and propagation by way of enhanced mass transfer and interphase mixing between the phases and also can lower the requirement of the severity of the operating conditions in terms of temperature and pressure.

Ultrasound irradiation causes cavitation of bubbles near the phase boundary between the alcohol and oil phases. The asymmetric collapse of the cavitation bubbles disrupts the phase boundry. The cavitation may also lead to a localized increase in temperature at the phase boundary, enhancing the transesterification reaction. Neither additional agitation nor heating is required to produce biodiesel by ultrasound technique.

The present work aims at exploiting continuous production of biodiesel using ultrasonic energy with the aim of reducing the processing cost. Trans-esterification of non-edible jatropha vegetable oil with methanol in the presence of alkaline catalyst potassium hydroxide has been investigated under the influence of ultrasonic irradiation at 24 kHz frequency and 200 W as power input. The effect of different operating parameters such as ultrasonic power dissipation, catalyst concentration, methanol to oil molar ratio, flow rate of oil and residence time have been studied for process optimization. Excellent conversions of oil to biodiesel were obtained for all conditions studied in this work. Conversion levels of more than 95% have been observed with the use of continuous reactor in one pass using ultrasonic energy in less than 1 min residence time under ambient operating conditions.

Experimental

Materials and methods

In the present study non-edible jatropha oil was used containing about 14% free fatty acid (FFA). The alcohol and catalyst used in the present study were methanol and KOH. Methanol (99.8%) and KOH (98%) were purchased from Merck Specialities and Qualigens Fine Chemicals, respectively.

Experimental set-up

A Schematic diagram of the ultrasonic continuous flow cell reactor is shown in Figure 1. Vessel 1 was used for storage of vegetable oil and vessel 3 was used for storage of dissolved pottasium hydroxide in methanol. Both reactants were pumped through reciprocating pumps (2 and 4) into the ultrasonic flow cell reactor through a Y- joint at the bottom of reactor. Product mixture was continuously removed from the top of the reactor. The reactor consists of the ultrasonic generator having 24 kHz frequency with total supplied power input 200W. The ultrasound system has the facility to change the amplitude of the power ranging from 30-100%. The actual power dissipated in the system was computed through the software and system is assisted by computer for its monitoring and control. The volume of the reactor was 14 ml and diameter of the horn used was 14 mm. The flow cell reactor was equipped with a jacket, which permitted the recirculation ofchilled water from a constant-temperature bath and temperature of the reaction mixture was maintained within ± 1^0C. The measurement of temperature of the reaction mixture was done at the exit of the reactor with the help of the thermocouple.

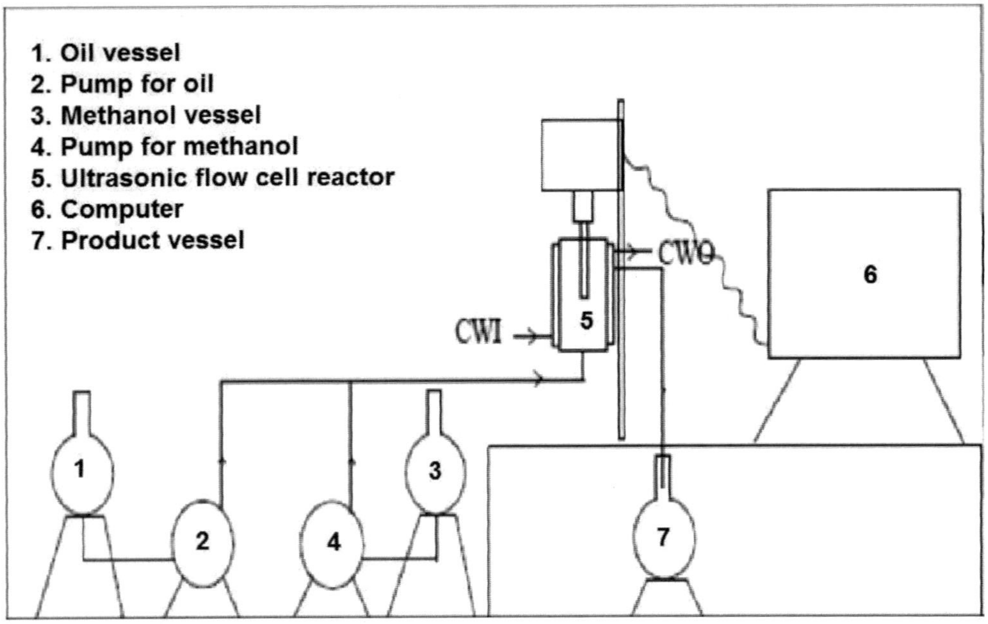

Fig. 1. A systematic diagram of experimental set-up of ultrasonic flow cell reactor

Experimental procedures

The two pumps of the reactants were adjusted at the desired flow rates to pass both the reactants into the flow cell reactor. Ultrasound generator was then switched on. Samples were collected for 2 minute after 1 minute, 3 minute and 5 minute time of start of ultrasound to study the steady state conditions of the system. Collected product was neutralized with acid and kept in the separating funnel for two hours to separate the phases. Bottom layer consists of glycerol rich phase and the upper layer consists of biodiesel rich phase. Bottom layer phase is separated with the help of gravity from the bottom of the separating funnel. The upper layer biodiesel rich phase was given a treatment of water washing, so that the catalyst and methanol present in this layer are separated. After the water washing, the sample was distilled in rotavapor to remove the trace amount of moisture present in the biodiesel. The samples were kept at 18^0C till further analysis.

Analysis

The compositional analysis of biodiesel product samples was done using a Gas Chromatograph of Chemito CERES 800 plus equipped with a flame ionization detector and a capillary column of VF 5 ht, 15m, 0.32mm, 0.10 micrometer. The GC oven was kept at 80^0C for 5 min, heated at 10^0C/min up to 450^0C, where it was kept for 1 min, and a total analytical time was 40 min. The carrier gas was nitrogen. The analysis of a sample by GC was carried out by injecting 1 micro liter of the sample solution into the GC. The formed methyl ester was identified by comparing its retention time to the retention time of standard methyl ester of fatty acid.

Result and discussion

The continuous flow reactor using ultrasonic energy is efficient and time saving for transesterification of vegetable oil with alcohol to produce biodiesel. Source of sound energy proved suitable for large-scale processing in continuous way of vegetable oil to biodiesel. Ultrasound assisted cavitation mainly affects the mass transfer rates and ensures uniform distribution of reactants as well as release high energy to get maximum conversion at shorter residence time in the continuous reactor. As we have discussed in previous section that apart from the cost of vegetable oil for lowering the cost of biodiesel the other main factors that influence the cost of production of biodiesel are energy consumption, the amount of methanol and catalyst. During this study of the transesterificaton reaction, the effects of various parameters such as the dissipation of ultrasonic power, catalyst concentration, molar ratio of methanol to oil, flow rate/residence time on the conversion have been investigated. The reaction samples were collected after 1, 3 and 5 minute after start of ultrasound to study the steady state of the system and it was found that the conversion was same for each sample. Thus, it can be said that the steady state was attained within one minute operation of the system. The optimum operating parameter for around 100% conversion in the present work were observed to be 6:1 mole ratio of methanol to oil, 1.25 wt% catalyst concentration and temperature 30°C and residence time less than one minute. The present work has clearly demonstrated the key role of continuous reactor using ultrasonic energy for biodiesel synthesis process.

Conclusion

The present work has clearly demonstrated the key role of continuous reactor using ultrasonic energy for biodiesel synthesis process. The continuous flow reactor using ultrasonic energy is efficient and time saving for transesterification of vegetable oil with alcohol to produce biodiesel.

(444a) Novel Catalysts and Reactor Concepts for Sustainable Processes: the Use of LCA Data in Process Design

Yaseneva, P.,
Lapkin, A., University of Cambridge
Originally presented on: 11/6/2013 8:30:00 - 8:55:00

Novel catalysts and reactor concepts for sustainable processes: the use of LCA data in process design

Polina Yaseneva and Alexei Lapkin

Department of Chemical Engineering and Biotechnology, University of Cambridge, Cambridge, CB2 3RA, United Kingdom

Abstract

We present a multi-step assessment framework, which was developed to guide decision making process in a process development context. Reaction mass intensity metric, gate-to-gate flowsheet analysis and LCA are used at appropriate stage gates. Specifically results of LCA assessment will be discussed for the case studies of a novel catalyst for water purification process, an intensive process of a pharmaceutical API synthesis and a novel organometallic catalytic process for intermediates synthesis.

Introduction

Assessment of sustainability of novel processes for decision-making purposes is a complex task, as frequently uncertainty of the data precludes the use of full LCA methodology, whereas proxy methods such as mass or energy metrics provide incomplete information. At the same time, information about sustainability of novel technologies is now routinely requested even within research projects. Thus, most EU projects in the area of chemical technologies are accompanied by LCA studies.

Within a large integrated project SYNFLOW (www.synflow.eu) we have developed a three-step approach, using reaction mass efficiency metric for selection of chemical routes, gate-to-gate mass and energy metrics for assessment on flow sheet options and finally LCA for evaluation of the demonstration case studies. These steps coincide with stage-gating within a project: confidence in chemistry to be taken up for process development, evaluation of process options with a focus on clean technologies (novel solvents, process intensification) and finally the detailed analysis of complete process at the stage of demonstration with LCA, targeting detailed flowsheet optimisation.

Here we present results of LCA studies for three specific examples. We show LCA of a novel carbon nanotube-based catalyst for water purification, focusing on comparison of the manufacture of the catalyst, a study of a novel flow-process of stoichiometric reduction of artemisinin, leading to manufacture of an important pharmaceutical API (focusing on comparison of batch vs continuous process) and finally a study of a novel catalytic Buchwald-Hartwig amination under overall flow conditions.

Results and Discussion

Carbon nanotube catalyst for water purification

A highly active catalyst was developed for reduction of an emergent aqueous pollutant, bromate, based on 0.3 %wt Pd deposited onto carbon nanotubes, grown on sintered metal fibre. The TOF of the catalyst was found to be an order of magnitude higher than that of a conventional catalyst based on alumina support (Pd/Al_2O_3). However, there remained the question whether high-temperature CVD process used in the manufacture of carbon nanotubes resulted in a cleaner overall process. The LCA study focused on the manufacture of the two catalysts. In this specific case all environmental impacts for the new catalysts were lower. Detailed analysis of contributions of the individual manufacturing steps to various impacts, e.g. analysis of contributions to cumulative energy demand shown in Figure 1, allows to reveal further options for optimisation of the novel catalyst manufacture [1].

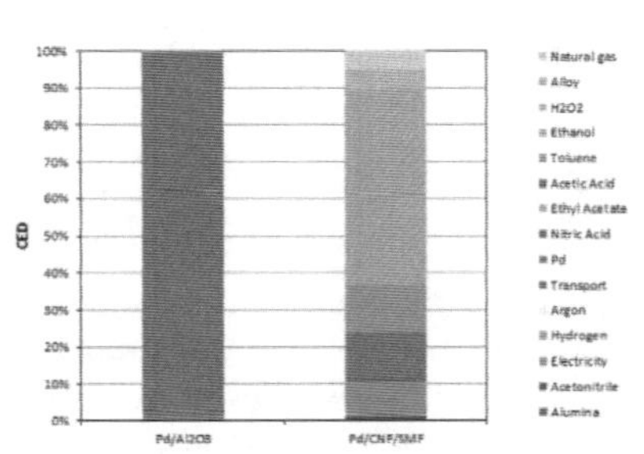

Figure 1. Individual contributions to cumulative energy demand (CED) of the manufacture of the two catalysts.

Artemisinin reduction under intensive flow conditions

We have developed a flow process for reduction of artemisinin to overcome the limitations of a conventional batch process [2]. In the current study we extended the reaction to the final step in the manufacture of the antimalarial API, but also performed an LCA study to compare the novel flow process with the conventional batch reaction [3]. Analysis of LCA results revealed that the solvent used in the manufacture and storage of the reducing agent required for the flow process is having a massive negative impact, see Figure 2. It is becoming critically important to replace the solvent used in a particular stage of the overall process. In this case LCA is instrumental in identifying the critical stage of the overall process that requires optimisation.

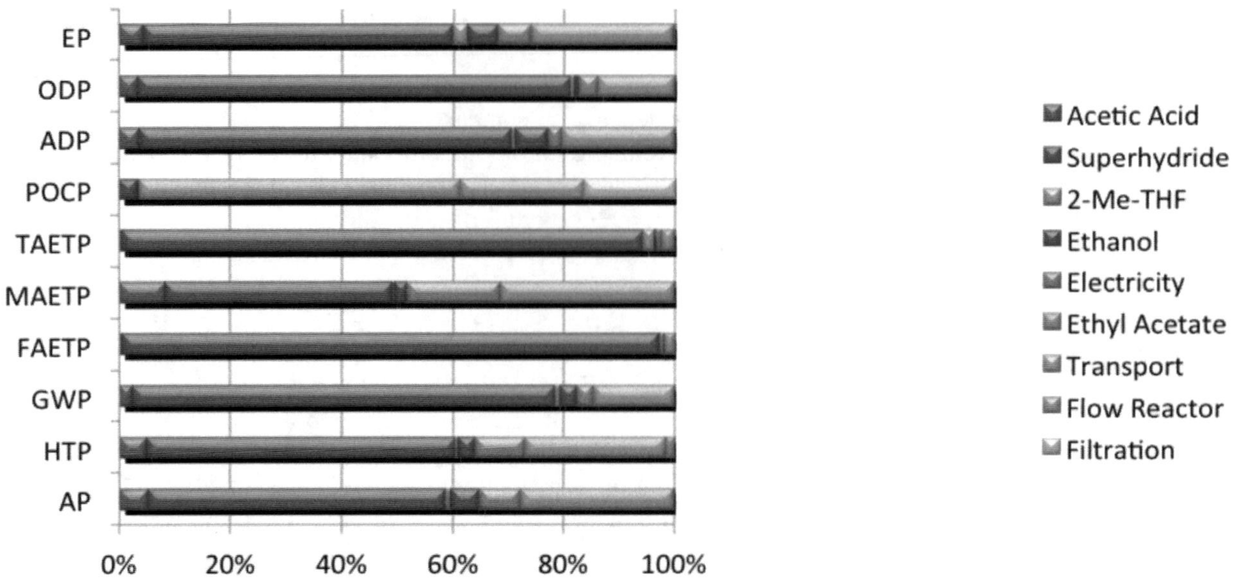

Figure 2. Individual contributions to environmental impacts in artemisinin reduction under flow conditions.

Buchwald-Hartwig amination in flow

The three-step decision making process is illustrated on the example of Buchwald-Hartwig amination. Reaction mass efficiency, process efficiency and LCA are evaluated for this reaction at the different stages in the process design. This approach complements the parallel chemistry-process design methodology being developed within Synflow project, ultimately aiming at a significant reduction of time required for commercialisation of novel technologies, which are also sustainable. Thus, for the initial reaction options we have used RME metric to evaluate the importance of solvent and catalyst recycle. The new process is currently being scaled-up to a demonstrator at an Invite facility of Bayer. We will show the results of stage-gating metrics and initial LCA data.

[1] P. Yaseneva et al., Reduction of bromates on carbon nanofibre supported structured Pd catalysts: experimental and life cycle assessment study, Submitted.

[2] X. Fan, V. Sans, P. Yaseneva, D. Plaza, J.M.J. Williams, A. Lapkin, Facile stoichiometric reductions in flow:

example of artemisinin, Org Proc Res Dev 16 (2012) 1039-1042.

[3] P. Yaseneva et al., Optimisation of the flow process of manufacture of artemether and its LCA study, Submitted.

Acknowledgement: This work was supported by the European Community's Seventh Framework Programme under project "SYNFLOW", NMP2-LA-2010-246461.

(444b) Techno-Economic and Environmental Assessment of Ionic Liquid-Based Biorefineries

Giarola, S., Imperial College London
Shah, N., Imperial College London
Originally presented on: 11/6/2013 8:55:00 - 9:20:00

Although the peak oil is still a debated issue, petroleum supply is expected to become ever more vulnerable of interruptions and current economic system reliance on fossils is exposing all economic sectors to unsustainable price fluctuations [1]. Concerns of maintaining an adequate supply of resources, such as fuels, food, raw materials for a growing population, are also raising a pressing interest for a change in our way to conceive energy and resource provision system [2]. A paradigm shift in the industrial sector from today's large fossils exploitation towards more sustainable practices is also advocated to face current trend of carbon dioxide emissions production, which may soon be approaching greenhouse gas (GHG) levels dangerously close to tipping points of an accelerated global warming [3]. Thus, the future challenge to address the growing world energy demand may be met as long as the supply of higher levels of resources and energy is achieved in a sustainable way promoting low-carbon technologies and renewables [4]. In this respect, the spread of biomass exploitation, as a unique source that can supply both energy and carbon for the production of the commodity chemicals and products, has the potentials to lead to the settlement of a new biomass-based economy [5-6].

Biomass is an abundant, renewable and carbon neutral resource, which is currently significantly exploited for biofuels production from edible components of food crops. However, the major challenge for a future bio-based economy would very likely rely on lignocellulosic feedstock exploitation, being more abundant and less expensive than starchy crops as well as avoiding any direct conflicts with the food chain [7]. Industrial biorefineries have been identified as the most promising route to the creation of a new domestic biobased industry, where biomass conversion processes into fuels, power, and chemicals are integrated through an efficient combination of technologies, to minimise the environmental footprints and ensure the sustainability of all products generated. By producing multiple products, a biorefinery can take advantage of the differences in biomass components and intermediates and maximise the value derived from the biomass feedstock [8]. However, cost-effective technologies for processing lignocellulosic materials into fuels and chemicals are not commercially available. In this context, a central role in determining the overall system profitability is played by the pretreatment unit, where biomass is deconstructed and the three main components (i.e. cellulose, hemicellulose and lignin) are made accessible to subsequent processes [9].

Since the study of Wilkes and Zaworotko [10], ionic liquids have become increasingly popular as reaction and extraction media. The property of ionic liquids which has led to them being classed as green solvents is their very low volatility, thus they do not emit volatile organic compounds and are suitable for reactions at atmospheric pressure. Recently, ionic liquids have begun to be investigated for biomass related processes, yet there are many technological applications and large commercial interests in the field [2]. The use of ionic liquids as lignocellulose deconstruction solvents is also extremely promising [11]. Among the potential applications, ionic liquid can be used to decrystallise cellulose and disrupt the lignin-hemicellulose network. The possibility of removing lignin with ionic liquids and recovering it as precursor for commercial products is also an interesting option [11]. However, deconstruction with ionic liquid will only be viable if some drawbacks are overcome, in particular the cost of ionic liquids. Process design and optimisation is required in order to have high solvent recovery and recycle, but also consumables use reduction, energy integration and waste minimisation. Economic aspects should also be coupled with the concept of ecological sustainability during the process optimisation strategy [12]. Process system engineering techniques play a major role in achieving all the above targets.

This work proposes an investigation of techno-economic performance of a biorefinery based on ionic liquid-pretreatment. The analysis, based on mathematical modelling and process design techniques, aims at identifying the key physical phenomena, the main cost centres affecting the process economics and proposing strategies to develop more efficient biochemical conversion routes for the biomass. This is performed along with the system environmental performance appraisal based on Life Cycle Assessment (LCA) techniques [13]. The final purpose is to provide a mathematical tool for an in-depth economic and environmental analysis of the ionic liquids fractionation, identifying the process enhancements alternatives and outlining potential synthetic routes to biocommodities.

References:

1. Lutz, C., U. Lehr, K. S. Wiebe (2012). Economic effects of peak oil. Energy Policy, 48, 829-834.
2. Tan, S. S. Y., D. R. MacFarlane (2009). Ionic liquids in Biomass Processing. Topics in current chemistry, 290, 311-340.
3. Foley, J. A. (2005). Tipping Points in the Tundra. Science, 310, 627-628.
4. International Energy Agency (2012). World Energy outlook. IEA, Paris (FR).
5. Clark, J. H., F. E. I. Deswarte, editors (2008). Introduction to Chemicals from Biomass Padstow: John Wiley & Sons, Ltd.
6. Ragauskas, A. J., C. K. Williams, B. H. Davison, G. Britovsek, J. Cairney, C. A. Eckert, et al. (2006). The Path Forward for Biofuels and Biomaterials. Science, 311, 484-489.
7. The Royal Society (2008). Sustainable biofuels: prospects and challenges. The Royal Society, London (UK). www.royalsociety.org
8. National Renewable Energy Laboratories (2009). Biomass Program. U.S. Department of Energy, Golden (CO).
9. Aden, A., M. Ruth, K. Ibsen, J. Jechura, K. Neeves, J. Sheehan, et al. (2002). Lignocellulosic biomass to ethanol process design and economics utilizing co-current dilute acid prehydrolysis and enzymatic hydrolysis for corn stover. Technical Report. National Renewable Energy Laboratory (NREL), Golden (CO).
10. Wilkes, J. S., M. J. Zaworotko (1992). Air and water stable 1-ethyl-3methylimidazolium based ionic liquids. Journal of the Chemical Society, Chemical Communications, 965-967.
11. Brandt, A., J. Gräsvik, J. P. Hallett, T. Welton (2013). Deconstruction of lignocellulosic biomass with ionic liquids. Green Chemistry, 15, 550-583.
12. Kralisch, D., D. Reinhardt, G. Kreisel (2007). Implementing objectives of sustainability into ionic liquids research development. Green Chemistry, 9, 1308-1318.
13. International Organisation for Standardisation (1997). ISO 14040: Environmental Management - Life Cycle Assessment - Principles and Framework. International Organisation for Standardisation: (ISO), Geneva.

(444c) Optimal Simultaneous Production of Bio-Ibutene and Bioethanol From Switchgrass

Martin, M., University of Salamanca
Grossmann, I. E., Carnegie Mellon University
Originally presented on: 11/6/2013 9:20:00 - 9:45:00

Ibutene is an important intermediate that has been typically produced from the cracking of the C4 fraction of crude oil. It is the basis for the production of a common additive to gasoline in search for a cleaner burning fuel, the methyl-tert-butyl ether (MTBE), as well as a monomer for the polymerization such as butyl rubber. Its importance can be somehow reflected in its price, around $2/kg. However, the dependency on the crude oil and the already limited availability due to the large number of applications has increased the need for new sources of this chemical. Ibutene has recently been used as intermediate for the production of oxygenated diesel substitutes from glycerol (Chen et al 2011, Vlad et al 2010, Martín & Grossmann 2013). In this application the main drawbacks are, on the one hand, the production of diesel substitutes using a fossil fuel based intermediate and, on the other hand, the economics since from the energy and water consumption standpoint its use for the enhanced production of diesel substitutives is competitive with the process that directly sells the glycerol as byproduct.

In this work, we propose the optimal flowsheet for the production of ibutene from switchgrass. A superstructure embedding a number of alternatives is proposed. Two technologies are considered for switchgrass pretreatment, dilute acid and ammonia fibre explosion (AFEX) so that the structure of the grass is broken down. Surface response models are developed for predicting the effect of the operating conditions on the yield of each pretreatment. Next, enzymatic hydrolysis follows any of the pretreaments to obtain fermentable sugars, mainly xylose and glucose. Ibutene is obtained by fermentation of the sugars. However xylose cannot be easily converted, and thus we also evaluate the possibility of using it to produce bioethanol. Next the ibutene has to be separated mainly from CO_2 for which, PSA or membrane separation are considered while in case ethanol is produced, a multi-effect distillation column followed by molecular sieves are used to dehydrate the ethanol. The problem is formulated as an MINLP with simultaneous optimization and heat integration based on Duran and Grossmann's model (1986). Finally, an economic evaluation is performed.

The most promising process involves the use of dilute acid pretreatment and membrane purification of the ibutene. The production of ibutene using renewable sources is feasible, and may be competitive for its use to generate diesel substitutes for glycerol etherification. However, the decision related to the production of ibutene alone or the simultaneous production of ibutene and ethanol depends on the prices for bioethanol and for switchgrass. For low and current biomass cost the production of ethanol ad ibutene is interesting for bioethanol process above $1/gal.

References:

Chen, P., Min, M., Chen, Y., Wang, L., Li, Y., Chen, Q., Wang, C., Wan, Y., Wang, X., Cheng, Y., Deng, S., Hennessy, K., Lin, X., Liu, Y., Wang, Y., Martinez, B., Ruan, R., 2009. Review of the biological and engineering aspects of algae to fuels approach. Int. J. Agric. Biol. Eng. 2, 1–30.

Duran, M.A.; Grossmann, I.E. (1986) Simultaneous optimization and heat integration of chemical processes. *AIChE, J.,* 32, 123-138

Martín, M., Grossmann, I.E. (2013) Simultaneous dynamic optimization and heat integration for the co-production of diesel substitutes: Biodiesel (FAME & FAEE) and glycerol ethers from algae oil. To be Submitted

Vlad, E., Bildea, C.S. and Bozga, G., 2010, Integrated design and control of glycerol etherification processes, Bull. Inst. Pol. Iasi, LVI (LX), 4, 139-148.

(444d) Synthesis and Design of Biorefinery Processing Networks With Uncertainty and Sustainability Analysis

Cheali, P., Technical University of Denmark
Gernaey, K. V., Technical University of Denmark (DTU)
Sin, G., Technical University of Denmark
Originally presented on: 11/6/2013 9:45:00 - 10:10:00

Synthesis and Design of Biorefinery Processing Networks with Uncertainty and Sustainability analysis

Peam Cheali, Krist V. Gernaey and Gürkan Sin

Department of Chemical and Biochemical Engineering, Søltofts Plads, Building 229, Technical University of Denmark, DK-2800 Lyngby, Denmark

Chemical industries usually rely on fossil based feedstock, which is a limited resource. In view of increasing energy demands and the negative environmental and climate effects related to the use of fossil based fuels, this motivates the development of new and more sustainable technologies for processing renewable feedstocks, with the aim of bridging the gap for fuel, chemical and material production. This project is focusing on biorefinery network design, in particular for early stage design and development studies.

Optimal biorefinery design is a challenging problem. It is a multi-objective decision-making problem not only with respect to technical and economic feasibility but also with respect to environmental impacts, sustainability constraints and limited availability & uncertainties of input data at the early design stage. It is therefore useful to develop a systematic methodology to support the evaluation of processes and the generation of sustainable alternatives for identifying the optimal processing routes. One particular challenge here is to include proactively sustainability analysis as part of the synthesis of biorefinery networks. Another challenge is the handling of several sources of uncertainties such as availability and composition of renewable feedstock, the technical performance of alternative processing technologies and the availability of future markets for biorefinery products, among others.

As part of earlier work in our research group, a systematic methodology to identify/generate optimal biorefineries was developed using the superstructure-based approach, and was implemented in a computer-aided framework. The methodology consists of tools and methods including databases, models, superstructure, and solution strategies to represent, describe and evaluate various combinations of processing networks. The optimization of the network is formulated as a mixed integer nonlinear programming type of problem and solved in GAMS. The methodology was applied for designing optimal biorefinery networks considering biochemical routes. Furthermore, the methodology has also been applied to soybean oil processing networks and industrial wastewater treatment networks. In the work presented here, the methodology for designing optimal biorefinery networks was expanded with inclusion of thermochemical routes, which means the scope and the size of the biorefinery network problem was extended at the level of the database, the models and the superstructure relevant for thermo-chemical conversion routes (the conversion of biomass feedstocks (corn stover, poplar wood) to fuels and chemicals (FT-gasoline, FT-diesel, bioethanol and higher alcohols) via thermal decomposition processes like pyrolysis and gasification).,

In this study, we extend the methodology, models and database by incorporating uncertainty and sustainability analysis as well. Appropriate ranges for uncertain parameters are identified with their correlation/covariance structure and latin hypercube sampling (LHS) is used to sample parameters from their respective domain of uncertainty. The parameter samples are then used as input for solving a deterministic and stochastic optimization problem. The sustainability analysis was performed in two ways: First, it was performed retrospectively to the optimal biorefinery network solution obtained after the MINLP by using an in-house software (SustainPRO) that employs ICHEME sustainability metrics. Secondly, the sustainability analysis was included proactively as part of the MINLP optimization problem that is performed to find the optimal processing path with respect to multi-criteria assessment including technical, economics and sustainability.

The expanded database and superstructure with uncertainty and sustainability analysis form a powerful process synthesis toolbox to be used in design of future biorefineries with multi-criteria evaluation (technical and economic feasibility, environmental impact and sustainability).

(444e) Ecosystems As Unit Operations for Innovative and Sustainable Process Design

Bakshi, B. R., The Ohio State University
Urban, R. A., The Ohio State University
Ziv, G.,
Originally presented on: 11/6/2013 10:10:00 - 10:35:00

Since their development in the 1920s, the concept of unit operations has played an important role in the design of chemical processes through the development of general models and software tools. With increasing importance of reducing the environmental impact of chemical processes, the concept of unit operations expanded to include processes for reducing pollutants in the streams leaving the process boundary. Systems engineering methods based on pinch analysis and optimization have also played an important role in enhancing the efficiency of manufacturing systems. For most chemical processes, it is not practical or economically feasible to completely eliminate material emissions or waste, and the chemical industry continues to be a major producer of hazardous waste and air pollutants, including carbon dioxide.

Ecosystems have the natural ability to treat a large variety of pollutants, and provide goods and services to manufacturing systems. For example, wetlands can provide fresh water by removing contamination by chemicals such as pesticides, oils, phenolic compounds, nitrates, etc. Trees provide water, biomass and oxygen, along with the ability to remove air pollutants such as carbon monoxide, particulate matter, sulfur oxides, carbon dioxide, etc. In fact, an unsustainable system may be defined as a system that produces waste in excess of the natural cleaning capacity of ecosystems. Conversely, sustainability requires that engineering be done within ecological constraints. Unfortunately, most current sustainable engineering methods, including life cycle assessment are not able to satisfy this requirement.

In this presentation, we will “rediscover” the bounty of nature and its importance in sustaining chemical processes. This will involve expanding the concept of unit operations by including ecological processes, to enable engineering design within ecological constraints. Process flowsheets will be designed as networks of technological and ecological systems. Monetary and environmental objectives will be considered, and results for objectives such as maximum profit, zero water, and zero carbon dioxide will be compared. Application to a biodiesel manufacturing process will show that the techno-ecological network is superior to a conventional technology-only process in terms of economic and environmental objectives. Case studies of other chemical processes will demonstrate the need to modify the chemistry or processing methods to permit full benefit of ecosystems. This can lead to new innovation that may not be discovered without including ecosystems among the design options. This approach of techno-ecological networks may be extended to larger spatial scales to result in unique ways of designing and operating supply chains and entire life cycles so that they stay within ecological constraints.

(444f) Process Analysis of Continuous Catalytic Gasification As a Waste to Energy Alternative

Tillie, C. F., Cleveland State University
Riedthaler, M. J., Cleveland State University
Adhikari, S., Cleveland State University
Reeves, S. A., Cleveland State University
Gatica, J. E., Cleveland State University
Originally presented on: 11/6/2013 10:35:00 - 11:00:00

Presently, space exploration waste is either carried onboard for the duration of the mission or discarded into a trash module which is sent back into Earth's atmosphere for destruction. On the other hand, with an estimated cost of $10,000 per pound, the cost of delivering mass to locations beyond low earth orbit (LEO) locations is prohibitive.

Since approximately 15 pounds of waste are produced daily by a crew of four during a mission, the intrinsic value of this waste can be greatly enhanced by waste to energy technologies. Moreover, conceptualizing waste gasification as an in-situ alternative for processing of trash is also an efficient route to control waste, while maintaining a healthy habitable environment, during long-duration missions (Santiago-Maldonado et al., 2010).

The gasification of long-chain polymers is hypothesized to follow a mechanism of four combined reactions —two parallel oxidation reactions which are kinetically controlled, accompanied by two series reaction where oxidation products (carbon dioxide and carbon monoxide) are converted into methane and hydrogen. Integrating this process into spaceflight missions would greatly reduce the onboard mass requirements, as these products can be used as propellant fuel. Further, under certain conditions the process could be also be adapted to produce oxygen and water, two essential life support species for space missions.

This paper explores scale-up alternatives to implement low-temperature continuous gasification processes as potential waste management processes. A kinetic model formulated from batch gasification experiments is integrated into a process simulator along with a detailed thermodynamic description.

Emphasis is placed on the preparatory treatment of the waste slurries, the reactor module, and equilibrium stages for recycling reactants and recovering the catalyst. Detailed characterization of low and mid fidelity simulants is presented.

(477a) Remediation of Diesel-Contaminated Soil With a Biostimulation and Bioaugmentation Joint Process

Li, X.,
Lu, D., Tsinghua University
Zheng, L., Tsinghua University
Originally presented on: 11/6/2013 12:30:00 - 12:44:00

Remediation of diesel-contaminated soil with a biostimulation and bioaugmentation joint process

Xiaosen Li, Diannan Lu and Zheng Liu*

Department of Chemical Engineering, Tsinghua University, Beijing 100084, China

lixs09@mails.tsinghua.edu.cnludiannan@mail.tsinghua.edu.cn

*Corresponding author: Phone: 86-10-6277-9876

Email: liuzheng@mail.tsinghua.edu.cn

A new remediation process that integrates biostimulation and bioaugmentation was proposed for diesel-contaminated soil. In addition to the determination of the residue diesel in soil, the total dehydrogenase activity, the microbial community changes in terms of composition and quantity of both microbial community and functional genes were probed by molecular biological techniques. In this research, five bacterial strains capable of degrading diesel were isolated from contaminated sites, in which Rem-E gave the highest degradation efficiency compared to the others and thus used as the additive strain for the bioaugmentation. As for biostimulation, monopotassium phosphate and ammonium sulfate were added to achieve an optimal C:N:P being determined as 100:1.25:1. After 10 day biosimulation, the content of diesel reduced from 2.7% to 1.6%. The joint process by biostimulation and bioaugmentation, however, gave a reduction of disel content from 2.7% to 0.6% in 40 days. Pyrosequencing for the total DNA extracted from the soil showed the evolution of the microbial community and qPCR revealed the change of the functional genes during the remediation process. The result revealed that coupling degrading strain and source of phosphorus and nitrogen led to the highest dehydrogenase activity and degradation rate, which also recover almost all the microbial community in the soil after the remediation. This work showed the great potential of the coupling technology composed of bioaugmentation and biostimulation for the remediation of the oil-contaminated soil. It also offers a viable route for the application of molecular biological techniques as enabling tools in the monitor, design, implementation and assessment of bioremediation practice.

(477b) Electrically Enhanced Contaminant Degradation in Sediment Caps

Zhang, R., Tianjin University of Technology

Originally presented on: 11/6/2013 12:44:00 - 12:58:00

In-situ capping is known as an attractive and cost-effective method for remediation of contaminated sediments. Sediment caps with bioremediation will ultimately significantly improve it performance, but reports about methods to enhance contaminant degradation in sediment caps still very little. Electrically enhanced bioactive barrier □BIOLANCE□is a new process for in situ remediation reducible compounds in soil or sediments. Chlorinated benzenes are significant contaminants to environment.

Many chlorinated organic compounds are persistent organic pollutants due to their resistance to biodegradation and chemical degradation. The objective of this study was to determine if electrodes placed in sediment and poised at several volts could create a redox gradient and provide electron acceptor/donor to stimulate chlorinated contaminant degradation, which will prepare for sediment caps application. The results demonstrated that choose graphite as electrodes can lead to sustainable evolution of hydrogen. Hydrogen evolution displayed zero-order kinetics with the time in the initial stages in different voltage. Even high voltage can produce more hydrogen, but it can not keep longer time because the overpotentials on electrode surfaces will prevent its function. In our experiments, 4V is the best choice. At this condition, the redox and pH change were detected. It is possible that reductive dechlorination could have been achieved with lower redox potentials in the presence of enough hydrogen electron donors through electrodes. Just while sulfate reducing or the other favorable electron acceptor may have reducing priority than chlorinated compounds. The experiments of T-cell capping biodechlorination use electrodes will be further studied.

(477c) pH-Dependent Transport of Energy By-Products in the Subsurface

Prigiobbe, V., University of Texas at Austin
Hesse, M., University of Texas at Austin
Bryant, S., University of Texas at Austin
Originally presented on: 11/6/2013 12:58:00 - 13:12:00

pH-dependent transport of energy by-products in the subsurface

Valentina Prigiobbe[a], Marc A. Hesse[b,c], and Steven L. Bryant[a,c]
[a]*Department of Petroleum and Geosystems Engineering*, [b]*Department of Geological Sciences*, [c]*Institute for Computational Engineering and Sciences, The University of Texas at Austin, Austin, Texas, U.S.A.*

Abstract
The transport of solutes through reactive porous media can be enhanced by the e ffect of pH-dependent adsorption/desorption processes. This eff ect can be particularly critical when the solutes, such as toxic metals and radionuclides, have hazardous impact on the environment and the health.
To study the e ffect of pH-dependent adsorption/desorption on the toxic metal and radionuclide transport, we have developed an analytical and a numerical model coupling transport and surface complexation for a single-phase incompressible fluid. We have analyzed a one-dimensional (1D) system of conservation laws to account of the pH-dependence and longitudinal dispersion and to gain a physical intuition of the phenomenon as well as a 2D system to include the eff ect of transverse dispersion and to investigate more realistic scenarios.
An important feature of these reactive transport problems is that one of the conservation laws is for the total hydrogen ($c_{ht} = c_h - c_{oh}$), also called acidity, which gives rise to an additional nonlinearity to the systems distinguishing them from other competitive adsorption/desorption problems where hydrogen is not involved.

Here, we present the results for two cases. A first one of pH-dependent transport of naturally-occurring toxic metals in shallow aquifers mobilized through a brine acidi

ed by carbon dioxide (CO_2) leaking from deep geological storage [1,2]. A second case of pH-dependent transport of adsorbing radionuclides accidentally discharged into the subsurface from temporally storage pools [3,4]. In both cases, we have analyzed the transport through a reactive porous medium composed of hydrophilic iron-oxide.
In the fi

rst case, the 1D solution of the intrusion of a CO_2-acidi

fied brine into a shallow aquifer has two retarded fronts separated by a spike of concentration of desorbed metal with a maximum much larger than the initial value. This is a classical behavior of pH-dependent adsorption, which occurs when the adsorption of the proton is more favorable than of the metal. In a 2D system, the structure of the plume resembles the 1D solution which, migrating into the aquifer with a maximum concentration much larger than the initial one, may present a health hazard.
In the second case, the 1D solution of the injection of an alkaline brine containing radionuclides into an aquifer consists of a retarded classical front and an unretarded non-classical pulse traveling at the average fluid velocity. The suitable initial and boundary conditions associated with the interaction of pH-dependent adsorption and the longitudinal dispersion are responsible for this phenomenon. In a 2D system in addition to the retarded front and the fast pulse, a fast spreading plume forms due to mixing induced by transverse dispersion. The unretarded pulse is a new transport phenomenon due to the interaction of pH-dependent adsorption and hydrodynamic dispersion, it leads to the migration of a contaminant with a velocity much faster than expected and it should be taken into account in the design of contaminated site remediation.

References

[1] Apps, J.A., Zheng, L., Zhang, Y., Xu, T., Birkholzer, J.T. (2010), Evaluation of potential changes in groundwater quality in response to CO_2 leakage from deep geologic storage. Transp Porous Med 82, 215-246.
[2] Kharaka, Y.K., Cole, D.R., Hovorka, S.D., Gunter, W.D., Knauss, K.G., and Freifeld, B.M. (2010), Gas-water-rock interactions in Frio formation following CO_2 injection: Implications for the storage of greenhouse gases in sedimentary basins. Geology 34, 577-580.
[3] Prigiobbe, V., Hesse, M. A., and Bryant, S. L. (2012a), Fast strontium transport induced by hydrodynamic dispersion and pH-dependent sorption. Geophys Res Lett 39 L18401.
[4] Prigiobbe, V., Hesse, M. A., and Bryant, S. L. (2012b), Anomalous reactive transport in the framework of the theory of chromatography. Transp Porous Med 91, 127-145.

(477d) Experimental Investigation Using a Contaminant-Monitoring Sytem in Storm Drains: A Dichotomous Preliminary Risk Assessment

Lackey, D., University of Alabama at Birmingham
Peters, R. W., University of Alabama at Birmingham
Originally presented on: 11/6/2013 13:12:00 - 13:26:00

As humans have built cities using concrete, asphalt, and large areas of impervious materials, they have potentially disturbed the normal hydrologic circle, and have learned that they must build storm drains to convey excess rainwater and prevent flooding. But these storm drains have a drawback in that they can also lead to pollution of natural environments if they carry contaminants such as antifreeze, fertilizers, paints, used motor oil, tossed pet waste, cigarette butts, and litter. Everything other than pure water is a potential contaminant that degrades water quality. Pollution sources entering storm drains include: automobile emissions, residential wastes, agricultural runoff, urban activities, construction runoff, and businesses [Hyperspective Studios, 2002].

In the summer of 2010, in conjunction with the UAB Occupational Health and Safety (OHS) Department, Dr. Robert W. Peters, and two of his graduate students conducted a project involving a risk assessment of contamination of runoff water flowing into storm drains on the UAB campus [Lackey *et al.*, 2010]. The project was successful enough to spin off into further studies but it had one key shortcoming: it considered only the effects of point sources on storm drains and largely ignored the effects of non-point sources. The details of that study, as well as the steps that will be taken to address that shortcoming, and what further studies the results will lead to, are listed below.

To begin the study, the UAB OH&S Department gave the civil engineering graduate research assistants a series of overhead maps of the UAB campus that have marked storm drain inlet locations from a previous study. The study took place over two main strategic areas on the UAB campus, one that contained the majority of storm drains on the western academic side of campus, and the other that contained the majority of the storm drains on the eastern side of campus which is more focused on medical research. The students were helped by the inlet locations provided on the maps by O&HS, but there were several differences between the locations given on each map and the actual locations of the inlets. Therefore, the project involved an update on the actual locations of each inlet as well as a risk assessment of the likelihood of dumping unwanted chemicals.

In the study, the students identified four main types of inlets:

1. Standard Covered Curbside: The type that contain a concrete or iron cover and engulf large volumes of water from adjacent roadways and parking lots during a rainstorm.
2. Uncovered Grate: These are often found in the middle of parking lots or on pedestrian

walkways. Occasionally, these are even found in grassy areas where irrigation is common.

3. Combination Curbside Grate: Essentially a combination of Types A and B, these are found in the same areas as Type A.
4. Elongated Grate: Found in the same types of areas as Type B, and extremely common at entrances of parking decks.

In addition to the types, each inlet was also assigned a risk assessment value of 0 to 3 based on the priority of its labeling. A color was assigned to each as well for easier readability:

0 (blue) – Marker already present on drain inlet

1 (green) – Low dumping risk

2 (yellow) – Moderate dumping risk

3 (red) – High dumping risk

Some of the criteria for determination of the risk as 1, 2, or 3 are as follows:

1. Default condition; generally assigned unless the inlet was near a probable pollution source such as a trash receptacle, dumpster, laboratory entrance, or parking space.
2. Assigned if the inlet was near a potential moderate pollution source, such as a laboratory entrance or trash can, or if the inlet was on or adjacent to a parking space.
3. Assigned if the inlet was next to a dumpster or other probable major pollution source, or two conditions or more that would result in "2" applied.

The major results of the study, in terms of number of inlets of each type and risk factor, are displayed below in Table 1 and Table 2.

Table 1. Summary of type and risk factor of inlets in Strategic Area 1 (Main Campus).

Area 1 – Main Campus

Risk Factor	Type A	Type B	Type C	Type D	TOTAL
Level 1	68	64	3	7	**142**
Level 2	11	20	1	0	**32**
Level 3	0	2	1	0	**3**
ALL	**79**	**86**	**5**	**7**	**177**

Table 2. Summary of type and risk factor of inlets in Strategic Area 2 (Medical Campus).

Area 2 – Medical Campus

Risk Factor	Type A	Type B	Type C	Type D	TOTAL
Level 1	112	49	0	20	**181**
Level 2	7	6	0	2	**15**

Level 3	1	4	0	3	**8**
ALL	**120**	**59**	**0**	**25**	**204**

In late 2012, the O&HS department decided that they not only want to install warning stencils on the UAB campus, but also sensors that would detect whenever harmful chemicals are present. The idea was expanded on to not only call for "artificial nose" chemical sensors but also a contaminant monitoring system that would map out the data from each sensor in real-time and send alerts to concerned parties whenever harmful chemicals are detected. Despite an initiative to cut costs as much as possible, the budget only calls for 50 of these "artifical noses". Therefore, priority needs to be assigned to the drains most in need of monitoring. When determining this, it turns out that anthropogenic point sources that result from dumping, while being the best criteria for whether or not to use a warning stencil, are not neccessarily the best criteria to decide whether a storm drain is a good candidate for a contaminant monitoring system, or at least not the only criteria. In order to complete the picture, non-point sources such as emissions from traffic must also be considered. O&HS has determined that 11th Avenue and University Boulevard are the roads on the UAB campus that have the highest traffic, and therefore the most at risk for contamination by traffic emissions. Some other adjacent roads have also been determined to have significant traffic.

Therefore, it would be best to expand the study to a two-phase, or dichotomous, risk assessment. The scale will be expanded from 1-5, and storm drains which are adjacent to high-traffic roads will have their risk factor raised by two, while storm drains which are adjacent to medium-traffic roads will have their risk factor raised by one. The new scale will become:

0: marked storm drain not on a high-or-medium-traffic road.

1: unmarked low point-source risk storm drain not on a high-or-medium-traffic road or marked storm drain on a medium-traffic road.

2: unmarked medium point-source risk storm drain not on a high-or-medium-traffic road, unmarked low point-source risk storm drain on a medium-traffic road, or marked storm drain on a high-traffic road.

3: unmarked high point-source risk storm drain not on a high-or-medium-traffic road, unmarked medium point-source risk storm drain on a medium-traffic road, or unmarked low point-source risk storm drain on a high-traffic road.

4: unmarked high point-source risk storm drain on a medium-traffic road, or unmarked medium point-source risk storm drain on a high-traffic road.

5: unmarked high point-source risk storm drain on a high-traffic road.

Storm drains which receive overall risk values of 5 will be the first to be monitored, followe.d by those with values of 4, and then those with values of 3, either until all 50 sensors have been placed or those of the 50 not meant to monitor "blank" base cases will be used. A "blank" would be one assigned a risk value of 0 or 1 that is used to determine if labeling or proximity to either point or non-point sources really does have an effect on contamination. And once the target drains are determined, the next phase of the project, which involves installing the sensors into the storm drains and beginning to analyze data generated by the contaminant-monitoring system, may proceed.

References Cited:

Hyperspective Studios, 2002. “Storm Drains”, website: http://protectingwater,com/storm-darins.html;

website accessed: October 16, 2012.

Lackey, D.A., A.N. Kajale, R.W. Peters, and J.D. Hagan, 2010. "The Effects of Labeling Storm Drains on the University of Alabama at Birmingham Campus", Poster paper presented at the Alabama Water Resources Conference, Orange Beach, Alabama (September 8-10).

(477e) Innovative Polymer Quantification and Management At a Large Scale Remediation Project

Grieco, S., O'Brien & Gere Engineers, Inc.
Redding, A.,
Lim, N.,
Originally presented on: 11/6/2013 13:26:00 - 13:40:00

Contaminated sediment remediation projects typically require a physical/chemical sediment pore-water treatment plant for metals and organics. In many cases, the sediments are thickened and dewatered for cost effective management and disposal. The polymer used for thickening can be problematic for downstream pore-water treatment. Specifically, this paper reports on a treatment system that experienced excessive "sticky" solids build-up in the clarifier and difficultly removing these solids through normal operations. The polymer also increased backwashing of Granular Activated Carbon (GAC), and decreased adsorption performance of Granular Activated Carbon (GAC).

In order to optimize thickening polymer management, this program identified the threshold of residual polymer that adversely impacted the pore-water treatment plant chemical precipitation & gravity settling unit operation. A mathematical relationship was established between a field-utilized China Clay Procedure and laboratory particle charge density (PSD) to establish an acceptable real-time simple procedure to validate the operating range and upper threshold for polymer dosing.

This program also quantified the impacts of residual polymer on the GAC. Equilibrium isotherms and rapid small scale column tests (RSSCTs) were conducted to confirm that the excess polymer was causing the observations in the full-scale GAC operations. Laboratory testing was used to quantify reduction of the bed volumes processed before both excess pressure drop and benzene breakthrough occurred. The program evaluated chemical types and dosages to dislodge and/or oxidize the excessive polymer that adhered to the carbon in the GAC beds.

The results of this testing were used in the field to optimize polymer dosing, which lowered operating costs and reduced system maintenance. The results also identified the impacts of GAC cleaning (both forward maintenance dosing as well as during backwash) on both pressure drop and adsorption capacity recovery.

(477f) Agile Multiple Water Network Designs for Soluble Contaminant Transport Identification and Site Remediation

Cai, T., Lamar University
Xu, Q., Lamar University
Originally presented on: 11/6/2013 13:40:00 - 13:54:00

Water network design (WND) is widely studied and employed in many industrial processes to minimize freshwater consumption and wastewater generation. Previous studies have been focusing almost exclusively on identifying one optimal WND for a studied process system. Actually, the quest for agile manufacturing by some multi-purpose industrial processes needs a comprehensive WND having the maximum operational flexibility to deal with various production requirements. Thus, the most desirable way for such agile operation is to couple various WNDs into one super design, and then through an effective control strategy to accomplish different production requests by switching among these WNDs. However, this has added the difficulty in contaminant transport identification once the network has been exposure to soluble hazards. In this paper, a novel methodology for the design of such a flexible super WND has been developed. It simultaneously considers the integration of multiple WNDs, network control switch based on different purposes and soluble contaminant transport identification in order to target timely site remediation. The objective of this optimization problem is a super WND coupled with all the recipe-based WNDs; the number of regular and three-way valves needed for the super WND; the optimal design of all of the valve placements in terms of the minimum total valve-switching times (TVST) to swap one pair of WNDs at a time; the control strategy (valve opening/closing strategy) to switch among WNDs in order to identify contaminant transport and reduce/terminate its further impact. The developed systematic methodology is general and thus applicable for many potential network integration problems. Case studies have demonstrated the efficacy of the development and its significance in the context of agile operation.

(477g) Modeling Electrokinetic-Soil Cleaning By Spatial Averaging

Tijaro-Rojas, R.,
Arce, P. E., Tennessee Technological University
Originally presented on: 11/6/2013 13:54:00 - 14:08:00

To predict the transport of ionic species into soils, such as in the acid mine drainages, a model will be implemented. An anisotropic porous medium and application of external forces as electric potential will be used in a multi-scale modeling approach. A minimum of key simplifications will be assumed, in order to capture the fundamental system behavior in model soils that mimic actual ones. Geometry of the control volume, medium properties, and intensity of electric potential constitute important variables to be considered. Since, the mass transport of the ionic species in soils takes place in a domain that displays different scales (molecular, microscopic, macroscopic and mesoscale), a multi-dimensional (spatial) averaging approach will be applied. Once, the spatial-averaged model become available, computational approaches will be used to illustrate the behavior of fundamental aspects. As a complementary part of this research, an experimental approach is being developed to guide and validate the predicted results.

(477h) Dynamics, Source Diagnosis, and Risk Assessment of Persistent Organic Pollutants in Water and Sediments of the Urban Aquatic Bodies

Kumar, R.,

Originally presented on: 11/6/2013 14:08:00 - 14:22:00

Abstract:

Organic compounds which are persistent, posses toxic characteristics, liable to bioaccumulate, prone to long range atmospheric transport and can cause adverse environmental and human health effects at locations near and far from their sources can be defined as persistent organic pollutants (POPs). Polycyclic aromatic hydrocarbons (PAHs) and organochlorine pesticides (OCPs) are the two important categories of POPs which are highly toxic to human beings, animals and plants. Polycyclic aromatic hydrocarbons are the product of any combustion process involving materials containing carbon and hydrogen namely coal, oil, petrol, wood etc. Emissions from the anthropogenic activities predominate, but some PAHs in the environment arise from natural combustion such as forest fires and volcanoes. PAHs occur naturally in crude oil and form a significant component of the petroleum products such as paints, creosote (used in wood preservation) and asphalt (used for road paving). There are some minor biogenic sources like plants, algae/phytoplankton and microorganisms and small amounts are formed by diagenesis in the lake sediments.

Organochlorine pesticides comprise a group of structurally diverse compounds with a large number of chlorine atoms. These pesticides have a tendency to bioaccumulate in the food chain and pose a great threat to the human health and the global environment. In spite of the ban on their use they are still being used in some of the developing countries, especially in the tropical regions, for health, agriculture and preservation purposes. Pesticides that are not bound in soils or taken up by the plants and animals can drain into rivers and lakes and move into the aquatic food chain.

The lakes and the rivers are two important types of surface water systems containing only 0.3% of the total global fresh water which is generally used for drinking purpose. These surface water resources, therefore, are precious and extremely sensitive and once degraded it would take immense efforts to revive them. In view of the above water and sediment samples were collected from two different types of urban surface water bodies of India. One of the water bodies is a lake system of Bhopal (Madhya Pradesh-a central state of India) and the other body is the stretch of river Yamuna at Delhi-the capital city of India. The aquatic systems under investigation are the typical urban aquatic systems and exist in the densely populated areas. Both of the systems are in different geographical locations. These surface water bodies being urban are expected to suffer with some similar types of anthropogenic activities. The various types of industrial, agricultural and domestic activities in the catchment area create different level of anthropogenic stress.

For the extraction of PAHs from water and sediment samples the method followed by Chen et al. [1] was adopted. The mobile phase was a gradient of acetonitrile and degassed water (50% acetonitrile held for 7 min; linear gradient to 80% acetonitrile in 7 to 20 min; 80% acetonitrile held till 25 min and linear gradient to 95% acetonitrile from 25 to 30 min) at a flow rate of 1.5 mL min^{-1}. The analysis was carried out at 254 nm and the individual PAH was identified and quantified by comparing the retention time and peak area with those of the standards. The procedure for extraction and cleanup of OCPs from both water and sediments is the same as outlined by Wan et al. [2]. The mobile phase was acetonitrile and water (70:30) with isocratic mode and a flow rate of 1.3 mL min^{-1}. The analysis of aldrin, endrin, pp'-DDT and α-endosulfan was performed at the wavelength 220, 228, 238 and 214 nm, respectively. The individual pesticide was identified and quantified on the basis of retention time and peak area. PAHs and OCPs were analyzed with HPLC equipped with a UV-detector (Agilent 1100, USA) and recorder. An octadecyl endcapped RP-C_{18} column (4.6 mm×250 mm, 5 μm particle size) was used.

The total PAHs (∑PAH) concentration in the water of the lakes and the river ranged from 4.85 to 18.2 ng/mL and 0.29 to 46.2 ng/mL, respectively. The profile of individual PAH in the water of both the aquatic systems is similar but different in sediments. The total PAHs concentration in sediments ranged from 952 to 5353 ng/g and 649 to 7910 ng/g for the lakes and river, respectively. The maximum pesticide concentrations in water and sediments of the lakes were 4.12 (aldrin), 2.26 (endrin), 6.17 (pp'-DDT), 5.98 ng/mL (α-endosulfan) and 167 (aldrin), 124 (endrin), 513 (pp'-DDT), 349 ng/g (α-endosulfan), respectively. The aldrin and endrin were absent at all the sites of the Yamuna. The maximum concentration of pp'-DDT and α-endosulfan in river water was 3.37 ng/mL and 3.55 ng/mL and in sediments 525 ng/g and 194 ng/g, respectively.

A perusal of the data points out that certain areas of the two aquatic systems are severely contaminated with PAHs and pesticides. The ∑PAH concentrations in water at some points of the lake system are above the lethal concentration and may cause acute toxicity to the exposed organisms. The data in the case of river indicate that both water and sediments are highly contaminated with PAHs. In both the cases the source of PAHs is mainly pyrogenic in nature. The potential toxicity due to total PAH concentrations will be rare in the lake system. The scenario is totally different in the case of river sediments where at most of the sites the toxic effects will be occasional. The apparent reasons for the deterioration of the said water bodies due to PAHs are dumping of untreated waste and a common practice of open burning of a variety of solid waste on the banks of the aquatic bodies. Apart from this, localized activity like plying of diesel operated motor boats is also contributing to the PAH pollution in the lake system. The major contributing activities responsible for PAHs contamination to the river through surface runoff and atmospheric transport are thermal power plants, heavy vehicular traffic load and domestic and industrial waste. A comparison of the two aquatic systems under study indicates that the ∑PAH concentrations in the river are about two times higher than those detected in the lakes. The PAH profiles in the water of the two systems are similar but the profiles in the sediments are different.

As far as the contamination due to pesticides is concerned aldrin is detected only at a few sites of the lake system and the other pesticides are observed only at sites surrounded by agricultural land. Endrin, pp'-DDT and α-endosulfan, wherever detected in water, are generally above the prescribed limit of acute toxicity. Aldrin and endrin are not detected in water and sediments of the river. Both the water and the sediments of the river are heavily contaminated with α-endosulfan and pp'-DDT. The biological impairments due endrin in the lakes and pp'-DDT in the river will be frequent as their levels in the sediments are above PEL. The accumulation of pesticides in the sediments may affect the natural bacterial composition and can lead to anaerobic conditions which may result in more toxic byproducts such as sulfide and ammonia.

References:

[1]. Chen, B., Xiaodong, X., Lizhong, Z., Jing, W., Yanzheng, G., Kun, Y., Xueyou, S. and

Baofeng, L., "Distributions of polycyclic aromatic hydrocarbons in surface waters, sediments and soils of Hangzhou City, China", Water Res. 38, 3558 (2004).

[2]. Wan, M.T., Kuo, J. and Pasternak, J., "Residues of endosulfan and other selected organochlorine pesticides in farm areas of the lower Fraser valley, British Columbia, Canada", J. Environ. Qual. 34, 1186 (2005).

(477i) Mass Transfer and Reaction Kinetics Limitations in Hazardous Waste (Organic and Metal) Treatment

Chawla, R., Howard University
Ighere, J., Howard University
Mlusu, J., Howard University
Farrell, A.,
Originally presented on: 11/6/2013 14:22:00 - 14:36:00

Improper storage and disposal practices have rendered hundreds of sites contaminated with hazardous substances. Some of the commonly found compounds include chlorinated organics such as trichloroethylene (TCE) and metals such as hexavalent chromium, Cr(VI), which occur simultaneously. This paper presents the feasibility of simultaneously treating TCE and Cr (VI), while addressing the problems related to (i) mass transfer limitations in chemical oxidation of TCE in soils, (ii) Effect of TCE oxidation on desired reduction of Cr (VI) to Cr (III), and (iii) removal of Cr(VI) using selected chelating agents.

Results of using alcohol cosolvents and household detergents for enhancing TCE solubility in water, thereby reducing mass transfer resistance in aqueous and soil systems will be presented. Redox reactions were studied using oxidizing agents, e.g. permanganate (MnO_4^-) for the remediation of TCE in the presence of Cr (VI) and using ferrous ions, Fe (II) for reduction of Cr (VI) to Cr(III) in the presence of TCE. In the redox reactions, TCE is oxidized to chloride, hydrogen, and carbon dioxide while permanganate is reduced to manganese dioxide (MnO_2). Similarly, Cr(VI) is reduced to Cr(III) and Fe(II) is oxidized to Fe(III). Batch experiments were performed to determine the reaction kinetics and extent of redox reactions of TCE and permanganate in the presence and absence of hexavalent chromium in aqueous systems. The reaction was determined to be second order overall, first order with respect to both TCE and MnO_4^- in the aqueous system. The TCE-Cr(VI) reaction was determined to be second-order overall, first order with respect to both Cr(VI) and TCE. The TCE-$KMnO_4$-Cr(VI) reaction at stoichiometric molar ratios (1:2:1), based on TCE reactions yielded 100% TCE oxidation. Cr(VI) was found to have a significant positive effect on the oxidation of TCE as it acted as an additional oxidant.

In the chelation study, batch experiments were performed to determine the degree and rate of chelation of Cr(VI) for three chelating agents- disodium dihydrogen ethylenediamine tetraacetate (Na_2H_2EDTA), ethylenediamine tetraacetic acid (H_4EDTA) and Nitrilotriacetic acid (NTA) Kinetic analysis indicated that the order of reaction was 1.32 with respect to Cr(VI) for Na_2H_2EDTA, 1.17 in the case of H_4EDTA and 2.38 for NTA. It was also found that the rate constants for the individual complexation reactions were 6.89×10^{-2} $[mg\ Cr(VI)/g]^{1-n}hr^{-1}$, 4.98×10^{-2} $[mg\ Cr(VI)/g]^{1-n}hr^{-1}$ and 7.25×10^{-2} $[mg\ Cr(VI)/g]^{1-n}hr^{-1}$ respectively.

(477j) A Simulation Of Weathering Of Colombian Crude Oils In The Colombian Caribbean Sea

Ramírez, J., Universidad Nacional de Colombia – Sede Medellín, Facultad de Minas, Bioprocesos y Flujos reactivos

Merlano, A., Universidad Nacional de Colombia – Sede Medellín, Facultad de Minas, Bioprocesos y Flujos reactivos

Guerrero-Gallego, J., Universidad Nacional de Colombia – Sede Medellín, Facultad de Minas, Grupo de investigación en Oceanografía e Ingeniería Costera - OCEANICOS

Peláez, D.,

Hernandez, J., Universidad Nacional de Colombia – Sede Medellín, Facultad de Minas, Grupo de investigación en Oceanografía e Ingeniería Costera - OCEANICOS

Osorio, A., Universidad Nacional de Colombia – Sede Medellín, Facultad de Minas, Grupo de investigación en Oceanografía e Ingeniería Costera - OCEANICOS

Molina, A., Universidad Nacional de Colombia – Sede Medellín, Facultad de Minas, Bioprocesos y Flujos reactivos

Originally presented on: 11/6/2013 14:36:00 - 14:50:00

A SIMULATION OF WEATHERING OF COLOMBIAN CRUDE OILS IN THE COLOMBIAN CARIBBEAN SEA

Juan Ramírez[1], Aura Merlano[1], Juan Guerrero-Gallego[2], Daniel Peláez[2], Juan Hernández[2], Andrés Osorio[2], Alejandro Molina[1,*]

[1] *Universidad Nacional de Colombia ? Sede Medellín, Facultad de Minas, Bioprocesos y Flujos reactivos, Medellín, Colombia*

[2] *Universidad Nacional de Colombia ? Sede Medellín, Facultad de Minas, Grupo de investigación en Oceanografía e Ingeniería Costera - OCEANICOS, Medellín, Colombia*

[*] *Corresponding author: amolinao@unal.edu.co*

A module, MEUN (*Módulo de Envejecimiento Universidad Nacional*), that describes the processes that occur due to the interaction, also known as weathering, of the crude in an oil spill with the atmosphere and the ocean, was developed. This module, programmed in FORTRAN, couples individual sub-models available in the literature to describe the evaporation, emulsification, dispersion and spreading processes that characterize the first week after an oil spill in the ocean. MEUN predicts as well the variation in density and viscosity over time as a result of the weathering processes.

In MEUN the evaporation submodel considered the pseudocomponent approach proposed by Jones, (1997). Knowledge of the True Boiling Point (TBP) curve and API gravity of the crude oil of interest, allowed the estimation, with Aspen Plus V7.2, of the thermodynamic properties (vapor pressure, mole fraction and mole volume) for each pseudocomponent. The weathering module includes the first-order kinetic expression proposed by Mackay et al., (1980) to model the rate of the emulsification process. The kind of emulsions formed (stable, mesostable, unstable or entrained water), the maximum water content of the emulsion and the evaporative thresholds to have a change in the type of the emulsion formed were interpolated from the experimental work developed by Fingas and Fieldhouse, (2004). MEUN couples the dispersion expression, developed by Delvigne and Sweeney, (1988), that takes into account the significant wave height and wave period to predict the amount of crude oil dispersed in the water column by the wave action. To predict the spreading rate of the oil slick, MEUN uses the equation developed by Lehr et al., (1984) based on observations in controlled oil spills. This expression considers the imbalance of forces acting on the oil slick and the effect of wind velocity to predict the spreading rate. The models developed by Lehr et al., (2002) and by Mackay et al., (1980), included in MEUN, predict the increase in density and viscosity, respectively, that results from the evaporation and emulsification processes.

All sub-model's constants were adapted to the specific requirements of Colombian crudes, particularly to Cusiana (°API 44.1) and Vasconia (°API 24.5). While the first one represents light crudes, the second one is an example of heavy oil. These two crudes have high production and require marine transport. The simulations results of the atmospheric model WRF (Weather Research and Forecasting Model), the wave model SWAN (Simulating WAve Nearshore) and the ocean model ROMS (Regional Ocean Modeling System) gave a typical ocean and atmospheric data of *Golfo de Morrosquillo* (latitude 09°29' N and longitude 75°46' W). Two scenarios, for high (4 m/s - 12 m/s) and low (0 m/s - 4 m/s) wind velocity, and the effect of time-averaging wind velocity, significant wave height and sea surface temperature were studied.

As an example of MEUN's predictions, Figure 1 shows the variation of the evaporated volume fraction with time after the spill for Cusiana and Vasconia oils. As expected, the lighter oil yields a higher evaporation fraction. Figure 1 also suggests that, while an average wind (discontinuous lines in Figure 1) yields reasonable predictions for a high wind condition, it significantly overestimate evaporation when the wind velocity (continuous lines in Figure 1) is low.

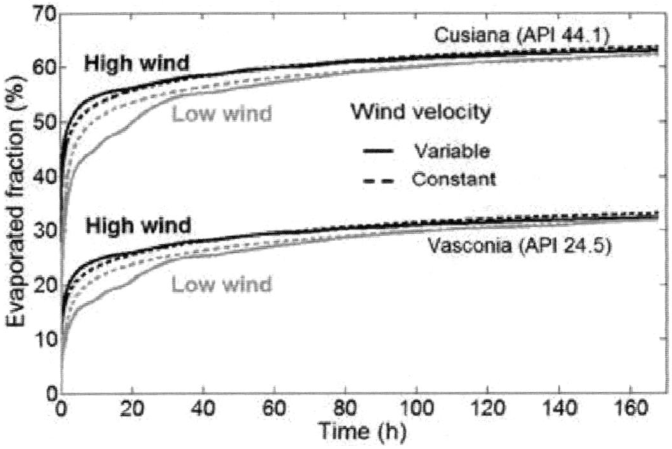

Figure 1. Predicted variation of the evaporated fraction for Cusiana and Vasconia crude oils after a spill in the Caribbean Sea under high and low wind velocity scenarios and variable and constant wind velocity.

Figure 2 summarizes MEUN's predictions as a plot of the volume fraction evaporated, dispersed and remaining on the sea surface. Because of differences in the physicochemical properties of the Colombian crude oils, Cusiana is more affected by evaporation and dispersion than Vasconia. According to MEUN, under high-wind conditions, a combination of evaporation and dispersion causes in Cusiana a decrease in the fraction of crude oil remaining on the ocean surface that is more than twice the decrease for Vasconia crude oil. While the results in Figure 1 and 2, as well as the fact that MEUN predicts an increase in viscosity of two to three orders of magnitude as the slick evaporates and the water makes an emulsion with the oil, agree with previous observations of weathering of oils in the ocean, a more complete validation of MEUN with experiments is currently under way using a wind tunnel and a rotating cylinder for evaporation and emulsion, respectively, at conditions typical of the Colombian Caribbean Sea.

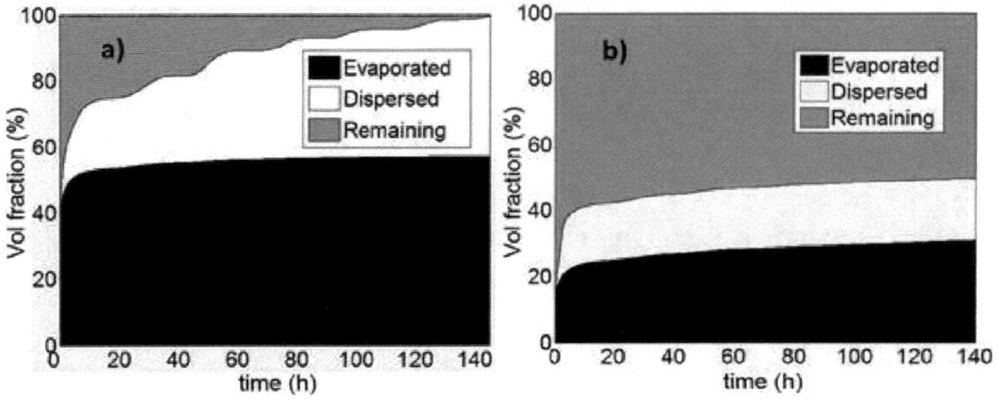

Figure 2. Distribution of the spilled crude oil between evaporated, dispersed and remaining on the ocean fractions for a) Cusiana and b) Vasconia oils. Predictions according to *MEUN*.

REFERENCES

Delvigne, G. A. L., & Sweeney, C. E. (1988). Natural dispersion of oil. *Oil and Chemical Pollution*, *4*(4), 281?310.

Fingas, M., & Fieldhouse, B. (2004). Formation of water-in-oil emulsions and application to oil spill modelling. *Journal of hazardous materials*, *107*(1-2), 37?50.

Jones, R. K. (1997). A simplified pseudocomponent oil evaporation model. *20th Arctic Marine Oilspill Program Technical Seminar.* (pp. 43?61). Ottawa, Ontario: Environment Canada.

Lehr, W. J., Fraga, R. J., Belen, M. S., & Cekirge, H. M. (1984). A new technique to estimate initial spill size using a modified fay-type spreading formula. *Marine Pollution Bulletin*, *15*(9), 326?329.

Lehr, W., Jones, R., Evans, M., Simecek-Beatty, D., & Overstreet, R. (2002). Revisions of the ADIOS oil spill model. *Environmental Modelling & Software*, *17*(2), 189?197.

Mackay, D., Buist, I., Mascarenhas, R., & Patterson, S. (1980). Oil Spill Processes and Models. Report No. EE/8. *Environment Canada* (p. 192). Ottawa, Ontario.

(477k) Increasing Strength of Thickened Oil Sand Tailings with Geopolymers

Xu, Y., Natural Resources Canada
Karimi, S.,
Originally presented on: 11/6/2013 14:50:00 - 15:04:00

The government of Alberta recently imposed new regulations requiring that oil sand tailings deposits must reach strength of 5 kPa one year after deposition. However, due to the limitations of gravitational settling and thickener operation, fine oil sand tailings can be thickened only to certain degree, and the strength may not meet the regulatory requirement. CanmetENERGY and Total E&P Canada (TEPCA) have conducted an investigation on the feasibility of using geopolymer additives to improve the strength of the deposit.

In this project, fly ash (type C) and cement were used as additives. Mature fine tailings (MFT) containing 35 wt% solids and thickened tailings (TT) with solids contents of 50 wt% were used as tailings material. Geopolymer was mixed with tailings and the treated tailings were distributed into 250-mL jars, some of which were closed, while others were left open. The solids content in the closed jars remained constant because no water evaporated, but increased in the open jars as the tailings dried in air at ambient conditions. The tailings strength was measured in each jar after a specified time.

These experiment results indicate that the addition of small amount of geopolymer increased the tailings strength significantly in both open and closed jars under the test conditions. Addition of about 5 wt% fly ash to TT increased the yield stress of the tailings to over 5 kPa after the solids content increased to around 60 wt% under natural drying. Similarly, adding around 5 wt% fly ash to the MFT increased the yield stress to over 5 kPa when the tailings were dried to about 55 wt% solids. Adding 3 wt% Portland cement also increased the yield stress of MFT or TT to more than 5 kPa when the solids concentration increased to 55 wt % for MFT and 65 wt% for TT. By comparison, untreated MFT or TT had to be dried to over 70 wt% and 85 wt%, respectively, in order for the yield stress to reach 5 kPa.

(677a) Study Of Hydroden Production By Photofermentation From Different Organic ACID As Substrate

Oliveira, T. V., Federal University of Uberlandia
Oliveira, F. S., Federal University of Uberlandia
Sousa, T. M., Federal University of Uberlandia
Ferreira, J. S., Federal University of Uberlandia
Batista, F. R. X., Federal University of Uberlandia
Cardoso, V. L., Federal University of Uberlandia
Originally presented on: 11/7/2013 12:30:00 - 12:55:00

Hydrogen presents itself as an alternative fuel in the search for clean energy. One of the biological routes for hydrogen production is the photofermentation by purple non-sulfur photosynthetic bacteria. In general, these cells are able to synthesize hydrogen from organic acids. In this study, the potential of the H_2 production of ***Rhodopseudomonas palustris and Rhodobacter capsulatus*** strains was evaluated. The influence of organic acids (butyric, malic, lactic, propionic and acetic) as substrate was also studied at fixed ration of C/N of 35 mM : 1 mM, using sodium glutamate as nitrogen source. The results obtained from the fractional factorial design 2^{5-1}, based on variance analysis (ANOVA), showed that only the variables statistically significant at 90% confidence were malic acid and lactic acid for ***Rhodopseudomonas palustris*** and malic acid, lactic acid and acetic acid for ***Rhodobacter capsulatus***. The results showed that ***Rhodopseudomonas palustris*** presented higher productivity (26.01 µmol/L.h) when malic, lactic and butyric acids were combined as substrate resulting in a conversion of 0.1151 (mol of H_2/mol substrate). For ***Rhodobacter capsulatus***, the higher hydrogen production was verified when malic, lactic and propionic acids were combined (19.09 µmol/L.h) as substrate resulting in a conversion efficiency of 0.0068 (mol of H_2/mol substrate).

Acknowledgments: FAPEMIG, CNPq, CAPES and Vale S.A. by financial support.

(677b) Influence Of pH, Ammonium and Iron On The Hydrogen Production By Microbial Consortium

RomÃ£o, B. B., Federal University of Uberlandia
Santos, J. G., Federal University of Uberlandia
Oliveira, F. S., Federal University of Uberlandia
Ferreira, J. S., Federal University of Uberlandia
Batista, F. R. X., Federal University of Uberlandia
Cardoso, V. L., Federal University of Uberlandia
Originally presented on: 11/7/2013 12:55:00 - 13:20:00

Biological hydrogen production processes are known to be friendly and consume less energy compared to hydrogen production through the conventional route, such as thermochemical and electrochemical methods. The aim of this study was to investigate the influence of pH (4 to 7), ammonium sulfate (0.3 to 2 g/L) and iron sulfate (0.1 to 0.5 g/L) concentrations in the dark fermentation for hydrogen production by microbial consortium, using a composite design (2^3), on the substrate conversion (mol hydrogen per mol consumed substrate) and hydrogen productivity. Lactose (20 g/L), from milk whey permeate, was used as the carbon source. The fermentation medium was complemented with the following nutrients (g/L): 3 KH_2PO4, 7 K_2HPO_4, 1 $MgSO_4$, 3 yeast extract and 0.5 meat extract. The findings showed that the culture condition that resulted in higher conversion (6.73 mol H_2/ mol consumed substrate) was obtained to pH of 7, iron sulfate (0.5g/L) and ammonium sulfate (2g/L). In addition, a high level of the hydrogen productivity (73.7 mmolH_2/L.d) was obtained in the same condition.

Acknowledgments: FAPEMIG, CNPq, CAPES and Vale S.A. by financial support.

(677c) Monitoring of Gaseous Sulfur Species in a Modified Sulfur-Iodine Thermochemical Water Splitting Cycle in a Hydrophobic Ionic Liquid

Caple, K., Oregon State University
Originally presented on: 11/7/2013 13:20:00 - 13:45:00

Monitoring of Gaseous Sulfur Species in a Modified Sulfur-Iodine Thermochemical Water Splitting Cycle in a Hydrophobic Ionic Liquid

Kevin Caple and A.F.T. Yokochi

The development of a clean, simple, and renewable energy source is becoming one of the more prevalent topics amongst scientists, researchers, and politicians alike. Processes involving solar, wind, nuclear, chemical, and a smattering of other combinations of these have been studied and explored from numerous angles. Of the various processes studied, thermal based water splitting may be one of the most popularly studied, yet least commercially viable options available on the market today, due to drastic inefficiencies.

Some of these inefficiencies in thermal water splitting can be mitigated through what has been coined "thermochemical cycles" which are processes that involve several sequential and circular chemical reactions to take water and produce by hydrogen and oxygen while regenerating the initial reactants. Of these thermochemical cycles, one of the most commonly studied and most straightforward is the Sulfur-Iodine Cycle, which uses sulfur dioxide, iodine and a large excess of water to produce iodic and sulfuric acid. The two acids are then further processed into their initial reagents, as well as the desired hydrogen product. However, the S-I cycle has several drawbacks, including expensive catalysts for the acid treatments, material compatibility, and separation difficulty due to low percentage azeotropes.

In an attempt to moderate these drawbacks, we have developed a modification of the S-I cycle utilizing a previously undesired side reaction. To further alleviate any material compatibility issues, rather than run the reaction in water, an imidazolium based ionic liquid was chosen as a solvent for the following series of reactions to occur:

$$I_2 + SO_2 + 2H_2O \rightarrow 2HI + H_2SO_4 \text{ (Bunsen Reaction, Low Temp)}$$

$$8HI + H_2SO_4 \rightarrow 4I_2 + H_2S + 4H_2O \text{ (Side Reaction Producing H2S, Low Temp)}$$

$$H_2SO_4 \rightarrow SO_3 + H_2O \text{ (Sulfuric Acid Gasification, \sim 200°C)}$$

$$SO_3 \rightarrow SO_2 + .5O_2 \text{ (Reformation of Sulfur Dioxide, Low Temp)}$$

$$H_2S + 2H_2O \rightarrow SO_2 + 3H_2 \text{ (Steam Reformation, High Temp)}$$

This work focuses on the low temperature aspect of the cycle, specifically on the first two reactions involving the Bunsen Reaction that produces the two strong acids and the hydrogen sulfide production reaction, studying kinetic parameters, the extent of the cyclic reaction, and the amount of hydrogen sulfide produced.

(677d) The Water Splitting Kinetics of Two-Step Solar Thermochemical Process With CeO2

Arifin, D., University of Colorado Boulder
McDaniel, A. H., Sandia National Laboratories
Weimer, A. W., University of Colorado at Boulder
Originally presented on: 11/7/2013 13:45:00 - 14:10:00

Projected growth in global population and continued industrialization of developing countries will increase total world energy consumption by 50% in the next three decades. This increased demand for energy will be largely met by increased fossil fuel consumption, thereby increasing anthropogenic carbon in the atmosphere and further fuelling geopolitical conflicts over control of dwindling energy resources. Producing H_2 renewably by splitting H_2O in a solar-based thermochemical process is an attractive solution to both of these 21st century problems.

Due to the high thermal stability and fast kinetics, ceria is a favored material for the solar thermochemical process. In order to maximize the solar-to-hydrogen conversion efficiency, an understanding of the mechanistic pathway and a detailed engineering model of the water splitting process is necessary.

In this work, we report the results of a numerical analysis that separates experimental artifacts from material-specific H_2 production behavior yielding high quality fits to experimentally determined H_2 production rates over the entire reaction time domain. The results of this study show that the water splitting process is a surface mediated process analogous to a homogenous reaction, with low apparent activation energy.

(677e) Oxidation/Reduction Kinetics of Fe Particles in a ZiO2/ SiO2 Fluidized Bed

Greek, B., University of Florida
AuYeung, N., University of Florida
Imponenti, L., University of Florida
Klausner, J., University of Florida
Originally presented on: 11/7/2013 14:10:00 - 14:35:00

Reaction kinetics are crucial in the design of thermochemical reactors for hydrogen production. This paper investigates the reaction kinetics of various mass percentages of elemental iron (Fe) in a mixture of nonreactive particles, consisting of powdered zirconia (ZiO_2) and silica (SiO_2), when under a stream of H_2O. In addition, the oxidized bed was reduced under a stream of carbon monoxide (CO), and the reaction kinetics were characterized.. The overall cycling characteristics of an Fe-based system were investigated.. Temperatures well above the sintering temperature of the powdered iron particles alone were investigated due to the sintering inhibition provided by the nonreactive ceramic particles. During oxidation, a stream of H_2O was passed through the powder bed at temperatures ranging from 550-950°C. Reduction was performed using a stream of carbon monoxide (CO) at temperatures from 800-1000°C, with the minimum to avoid carbon depositing (due to the Boudouard reaction) in the powder bed. Products were measured via a mass flow meter and mass spectrometer during oxidation and reduction, respectively. Powder bed weight was used to close the reactant/product mass balance. Average hydrogen production rates over the first minute of reaction of 0.1-0.2 SLPM per gram of reactive material were achieved. This work characterizes the reaction kinetics of a readily available iron particle reactant in Fe-Fe_3O_4 redox reactions, using water as an oxidizing agent and CO as a reducing agent.

(677f) Hydrogen Fueling Stations: Optimization of Fill-Up Process

Olmos, F., UCLA
Manousiouthakis, V., University of California Los Angeles, Los Angeles
Originally presented on: 11/7/2013 14:35:00 - 15:00:00

One key challenge faced today by hydrogen fueling stations is to provide safe and efficient fill-ups for hydrogen fuel cell cars. Based on a self-consistent thermodynamic model of hydrogen, a dynamic model of the fill-up process, and a process simulator, strategies for the optimal control of this process can be devised. The simulator of the fill-up process is subject to optimization and control studies were global optimality is sought in order to identify the optimal fill-up strategy.

(703a) Isolation, Identification and Characterization of High CO2 Tolerant Microalgae for Efficient CO2 Sequestration From Two Different Locations of India

Ghoshal, A. K., Indian Institute of Technology Guwahati

Originally presented on: 11/7/2013 12:30:00 - 12:55:00

Isolation, Identification and Characterization of high CO_2 tolerant microalgae for efficient CO_2 sequestration from two different locations of India.

Samarpita Basu[1], AbhijitSarma Roy[2], K.Mohanty[1, 3], A.K.Ghoshal*[1, 2, 3]

[1]Centre for Environment, Indian Institute of Technology Guwahati, Assam-781039, India

[2]Centre for Energy, Indian Institute of Technology Guwahati, Assam-781039, India

[3]Department of Chemical Engineering, Indian Institute of Technology Guwahati, Assam-781039, India

*Corresponding author (E-mail:aloke@iitg.ernet.in)

The accumulation of greenhouse gases like CO_2 in the earth's atmosphere as a result of human activities and industrialization causes serious consequences like global warming thereby increasing the temperatures of the surface air and subsurface ocean. Apart from other causes combustion of fossil fuel for energy is one of the most important for increased CO_2in the atmosphere. Emissions of CO_2 are likely to increase significantly in the near future due to increased energy demand from developing countries.Public concerns over severe environmental consequences that might result due to globally rising temperatures have brought forward the necessities to find out ways to reduce the emission of CO_2 to the atmosphere. Some of the strategies employed to reduce CO_2 emissions include the implementation of renewable energy sources as well as capturing of CO_2 to limit its release into the atmosphere. The current methods of capturing CO_2 include the use of physicochemical absorbents and mineral carbonation of $CO_{2.}$However, these methods have several drawbacks like high space requirements, high cost and potential CO_2 leakage over time. Therefore, a lot of attention is now being given to the biological fixation of CO_2. Photosynthetic microorganisms like microalgae and cyanobacteria use inorganic carbon for growth and hence can convert CO_2 from a point source into biomass. Among the photosynthetic microorganisms, microalgal capture of CO_2 is thought to be a feasible technology with energy efficient and environment friendly natureMostly microalgae are unicellular photosynthetic microorganisms that can fix CO_2 effectively from gaseous effluents (e.g., industrial flue gas) and have much higher photosynthetic efficiency and biomass productivity as compared to other energy crops Moreover, the microalgae biomass produced after CO_2 fixation can be put to several uses such as feed stock for a variety of biofuels, medications, cosmetics and nutritious foods, representing additional benefits from the microalgal CO_2 reduction process. The present study aimed to isolate high CO_2 and temperature tolerant microalgae capable of sequestering CO_2 from flue gas. Water samples were collected from two different locations, one from a freshwater body (26°11'15"N 91°45'4"E) of Guwahati in Assam, located in north eastern region of India and another from a freshwater body (22°55′22″N88°22′46″E) located in the vicinity of a thermal power plant in West Bengal, located in eastern region of India. Isolation of microalgae strains were carried out from the samples using BG 11(ATCC Medium 616) as the culture medium *.14 hours of light and 10 hours of dark regime was employed during growth using daylight-type 20 W fluorescent tubes and light intensity maintained was 5496 lux. In totality, three microalgal strains designated as SA1 (from Assam), SA2 and SA3 (from West Bengal) were isolated. After isolation of pure culture, CO_2 fixation studies were initiated using these strains using* 13.8 ± 1.5 % *CO_2 as the sole carbon source for microalgal cultivation.* 28s rRNA gene sequencing of SA1 followed by homology matching with NCBI GenBank database confirmed SA1 to be the green microalga, *Scenedesmus obliquus*. The 28s rRNA gene sequence was then submitted to NCBI Genbank and accession number (KC733762) was obtained. Growth kinetics and CO_2 fixation rates of the isolates were determined at varying culture conditions (nitrate and phosphate concentrations, photoperiods and temperatures). Studies were carried out using these strains at elevated cultivation temperatures of upto 40°C since flue gas has a temperature of around 40°C after the desulfurization process. Characterizations of the strains were done in scanning electron microscope as well as FTIR. FTIR gave an idea of the biochemical composition of the microalgal strains. Carbohydrate, protein, lipid and pigment content of the microalgal strains were determined to elucidate the potential of these strains for the production of value-added substances like biofuels which would offer additional benefits to the CO_2 sequestration process. The various growth kinetic parameters like maximum biomass value, maximum specific growth rate, overall biomass productivity and CO_2 fixation rates were estimated for the strain SA1.The microalgae strains SA2and SA3 isolated from West Bengal are currently being studied and DNA isolation and sequencing are being carried out. The gross calorific values of the strains were estimated to elucidate the lipid content of the microalgae strains and also to validate the use of the algal biomass for biofuel generation after the carbon sequestration process. SA1 showed enhanced CO_2 fixation rates and high biomass productivities which were higher than most of the related studies. At 13.8 ± 1.5%CO_2 and *25 °C,* maximum biomass (4.975 ± 0.003 gL^{-1}) and maximum CO_2 fixation rate (252.883 ± 0.361$mgL^{-1}d^{-1}$) were obtained. *At* elevated *temperature (40°C) and* 13.8 ± 1.5% *CO_2, maximum biomass(*0.883 ± 0.001 g L^{-1}) was obtained. *The* carbohydrate, protein, lipid, and chlorophyll content of the SA1 after CO_2 sequestration were 30.87 ± 0.64%, 9.48 ± 1.65%, 33.04 ± 0.46 and 6.03 ± 0.19% respectively, which were higher than previous reports. The Gross calorific value (GCV) was 23.02 MJ/Kg for SA1 after CO_2 sequestration which was significantly higher than most of the previous reports.

Keywords: Global warming, microalgae, CO_2 fixation, flue gas, *Scenedesmus obliquus*

(703b) Complex Energy System Which Combined Photovoltaic Power Generation With Electric Industrial Vehicle-Case Study of Maniwa SMART Factory-

Shimizu, A., Okayama Prefectural University
Notoji, Y., Graduate school of Okayama Prefectural University
Nakagawa, T., Graduate school of Okayama Prefectural University
Originally presented on: 11/7/2013 12:55:00 - 13:20:00

SMART factory project has been developed under a grant from the Japanese government at Maniwa biomass base in Okayama Prefecture. In this developmental study, the following two research objects have been addressed. One is to build a new technology concerning manufacturing of nanofibers with high additional value from wood thinned from forests and unused wood parts from sawmills, and the other is to develop a new energy system and the elemental technologies which are provided renewable energy to meet the energy demand at the new factory producing nanofibers. The research target is to reduce economically CO_2 emissions of the entire factory by introducing these technologies.

A present factory which is built at Maniwa base has been producing wood chips from forest thinnings and two kinds of energies as light oil and electric power are consumed. Electric power has been used as the power of chip manufacturing equipment mainly. Light oil has been consumed as fuel for forklift trucks, shovel loaders and others. Electric power consumption is caused by 60% of CO_2 emissions from the factory and the remaining CO_2 emissions are due to light oil. Therefore, only the reduction usage of electric power is not enough to reduce significantly the CO_2 emissions of the factory. So, it is necessary to reduce the total consumption of electric power and light oil. In previous study, as a way to reduce electric power and fuels of vehicle simultaneously, the authors have been proposed a method according combined system of photovoltaic power generation (PV) and electric vehicle (EV).

In this paper, the new energy system has studied that industrial vehicle such as forklift trucks become exchanged to drive electricity which is supplied by PV, and PV power fluctuation become leveled using the storage battery in the industrial vehicle.

As a result, it is able to reduce CO_2 emissions by 25% according to introduce the new energy system that forklift trucks are driven by PV power which is charged directly to the battery in forklift trucks and is supplied to the factory if the surplus electricity is generated. In addition, it has a great possibility to reduce CO_2 emissions by 50% according to introduce an advanced system that shovel loaders and grapples will be driven by PV as the same as forklift trucks.

(703c) Multi-Objective Optimization of Bioelectricity Supply Chain With Life Cycle Assessment and Social Impact Analysis

Slivinsky, M., Northwestern University

Yue, D., Northwestern University

You, F., Northwestern University

Originally presented on: 11/7/2013 13:20:00 - 13:45:00

Multi-Objective Optimization of Bioelectricity Supply Chain with Life Cycle Assessment and Social Impact Analysis

Maxim Slivinsky, Dajun Yue, Fengqi You*

Submitted to session 23C01 Sustainable Electricity: Generation and Storage

Abstract

There is a strong drive for exploration of bioelectricity production, which is rooted in the increasing costs and limited availability of fossil fuels as well as the desire for greater energy security. Bioelectricity provides a promising alternative to fossil fuel power generation due to the compatibility with power generation infrastructure, reduction in greenhouse gas emissions, and associated job creation. The existing power infrastructure, including power grid, steam turbines, etc., can be utilized in the production of bioelectricity given some equipment modifications at the plant level and in pre-processing facilities [1]. This results in a relatively low capital investment opportunity to produce electricity from biomass feedstocks. This work approaches the supply-chain problem using three objectives: economic, environmental, and social. The economic objective is minimization of total annualized costs, the environmental objective seeks to minimize the life cycle greenhouse gas emissions (GHG), and the economic objective is to maximize total accrued jobs.

In this work a multi-objective mixed-integer linear programming problem (MILP) is developed to model the supply-chain optimization of bioelectricity production from feedstocks such as corn stover, forest residues, and switchgrass. This MILP is solved using the ε-constraint method, where the Pareto-optimal solutions are determined by solving MILP sub-problems with varying values of ε [2]. A key constraint in the formulation is the minimum fraction of annual electricity produced from biomass for a given region. The MILP is analyzed through different scenarios corresponding to this constraint parameter. Economic multipliers are taken from the Jobs and Economic Development Impact (JEDI) model to determine the social impact on the region [3]. The total accrued employment consists of direct employment (e.g. construction), indirect employment (jobs in upstream supply chain), and induced employment (jobs from money spent locally). Since JEDI has data for only two boiler technologies, this formulation looks at circulating fluidized bed and stoker boilers. The economic objective takes into account feedstock and utilities costs, transportation costs, construction costs, electricity demand by region, and processing limits. The environmental objective uses life cycle analysis (LCA), considering gate-to-gate environmental impact analysis of the preprocessing facilities, transportation, and capacity expansions. GWP is calculated using the 100 year timeframe per the Kyoto Protocol.

This problem is evaluated through a county-level case study for Illinois, which consists of 102 counties. County data is used to determine geographic biomass availability, existing power plant locations, and electricity demand. The problem is solved to determine the optimal locations and capacities of preprocessing facilities and biomass technology expansions, as well as the network design and technologies used [4, 5]. The Pareto solutions to this MILP reveal the tradeoffs among the decision variables under economic, environmental, and social objectives. The scenario analysis shows the economies of scale associated with replacing a fraction of power generation with bioelectricity.

References

[1] Environmental Protection Agency [EPA], Combined Heat and Power Partnership (2007, September). ?Biomass Combined Heat and Power Catalog of Technologies?, http://www.epa.gov/chp/documents /biomass_chp_catalog.pdf

[2] B. H. Gebreslassie, Y. Yao, and F. You, "Design under uncertainty of hydrocarbon biorefinery supply chains: Multiobjective stochastic programming models, decomposition algorithm, and a Comparison between CVaR and downside risk," *AIChE Journal,* vol. 58, pp. 2155-2179, 2012.

[3] National Renewable Energy Laboratory [NREL], Jobs and Economic Development Impact Models [JEDI] (2012, September), http://www.nrel.gov/analysis/jedi/

[4] F. Q. You, L. Tao, D. J. Graziano, and S. W. Snyder, "Optimal design of sustainable cellulosic biofuel supply chains: Multiobjective optimization coupled with life cycle assessment and input-output analysis," *Aiche Journal,* vol. 58, pp. 1157-1180, Apr 2012.

[5] You, F., & Wang, B. (2011). Life Cycle Optimization of Biomass-to-Liquids Supply Chains with Distributed-Centralized Processing Networks. *Industrial & Engineering Chemistry Research*, 50, 10102?10127

* Corresponding author. Email: you@northwestern.edu

(703d) Techno-Economic Modeling and Analysis of Energy Storage Options for a Self-Sustained Solar PV Array

Peterson, B. A.,
Smith, J. D., Idaho National Laboratory
Originally presented on: 11/7/2013 13:45:00 - 14:10:00

Renewable energies, such as solar and wind, are an ever-growing part of energy consumption world-wide. The primary push to grow renewable energy production is two-fold: reducing carbon dioxide emissions, thus moderating the effects of increased atmospheric levels of the gas, and reducing dependence on a diminishing fossil fuel supply. However, renewable energies certainly have issues, chiefly their intermittent and non-dispatchable nature. This is to say that solar power is only available when the sun shines, and wind is only available when the wind blows. This makes supplying electricity to a given demand load impossible for stand-alone renewable energy sources, limiting the financial incentive to invest in renewable energies. Intermittency also causes frequency and voltage instability for the grid when renewables are integrated into it. Energy storage is integral to the viability of renewable energy as it allows for load-smoothing and frequency regulation. Currently, no one or two obvious energy storage technologies exist to economically meet the requirements of meeting demand with renewable energy.

There are numerous ways in which energy can be stored: chemically, electrochemically, mechanically, and electrostatically. A model is proposed to study the economic feasibility of several energy storage options available today for a self-sustained solar PV array. Hourly demand data and hourly solar generation were used to determine constraints on the studied energy storage component. Factors including system capital cost, system operating cost, various replacement cost, and energy storage replacement times were used to calculate a net present value of the energy storage options for different demand curves. The model was used to examine the full lifetime of the PV array, typically 25 years. The most promising energy storage technologies for financial feasibility of a generalized solar PV array are sulfur-sodium batteries and carbon-enhanced lead acid batteries, depending on the scale, timeframe, and location of the energy storage.

(703e) Solar Electricity Generation System With Combined Energy Storage and Steam Generation

Xu, R., Texas Tech University
Wiesner, T., Texas Tech University
Originally presented on: 11/7/2013 14:10:00 - 14:35:00

Solar steam power plants for electricity generation were launched commercially in 1984. These plants are known as SEGs projects. The oil as the heat transfer fluid (HTF) was pumped through solar field to capture solar energy. However, the oil was unable to maintain the temperature during off-sun periods. Thermal energy storage (TES) is useful to increase the solar multiple. Molten salt is usually employed as a stable energy storage medium. Two-tank and single-tank construction have been employed for parabolic trough-based solar steam power plants. In our paper, we propose an improved thermal storage concept aiming at increasing the operation hours and reducing construction cost. Our design is shown in the figure below(Figure 1,right), which incorporates the steam evaporator inside the single thermal storage tank. Aspen HYSYS is used to dynamically simulate the process, analyze the steam generation efficiency and compare with the single-tank base case(Figure 1, left).

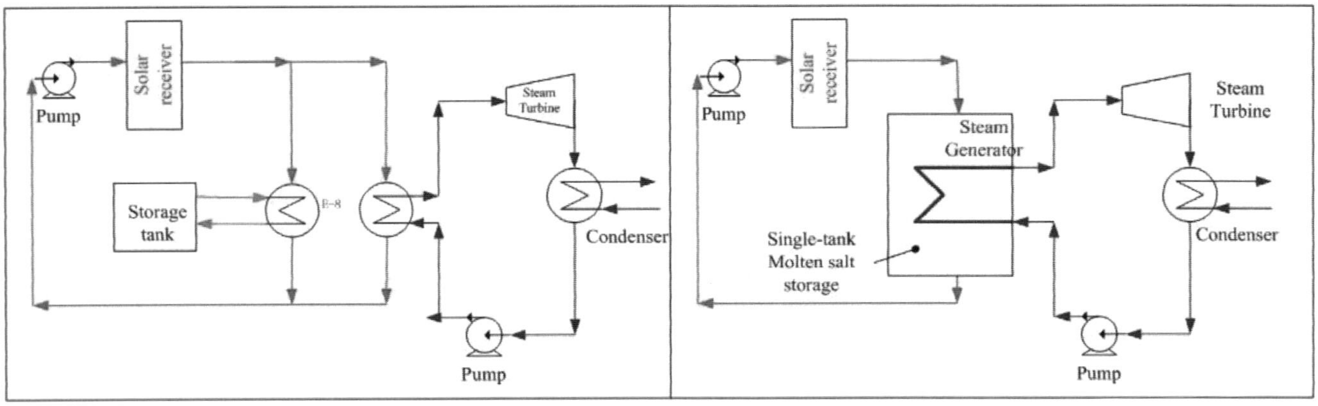

Figure 1 Thermal storage base case and the improved design process

(703f) Thin-Film Composite Membranes for Osmotic Power Generation

Li, X., National University of Singapore
Chung, T. S., National University of Singapore
Ong, R. C., National University of Singapore
Originally presented on: 11/7/2013 14:35:00 - 15:00:00

Thin-Film Composite Membranes for Osmotic Power Generation

Xue LI[a], Tai-Shung CHUNG-[a, b,*]

[a]NUS Graduate School for Integrative Science and Engineering, National University of Singapore, Singapore 117456, Singapore

[b]Department of Chemical & Biomolecular Engineering, National University of Singapore, Singapore 117576, Singapore *Corresponding Author's E-mail: chencts@nus.edu.sg **Keywords:** Osmotic power generation; Thin-film composite membranes; Pressure retarded osmosis. **Abstract**

Pressure retarded osmosis (PRO) is a membrane-based technology that harvests the energy of mixing two solutions from different salinity and converts it to electricity with the aid of a semi-permeable PRO membrane and a hydro-turbine. In addition to utilize seawater and river water as the feed pair, the PRO process can use highly concentrated RO retentate and recycled municipal wastewater as the feed pair for power generation. By doing so, not only can the environmental issue such as the disposal of RO retentate be solved, but also produce much higher osmotic energy. Therefore, the development of PRO membranes for osmotic power generation will have repercussion effects on the production of both clean energy and clean water.

As the heart of PRO processes, more studies on PRO membranes are expected to further accelerate the process in the near future. The thin-film composite (TFC) membranes, a commonly used PRO membrane nowadays, consist of a TFC layer of cross-linked polymers which acts as the primary barrier to salt and is responsible for giving a reasonable selectivity. In PRO processes, the draw solutions are pressurized by high hydraulic pressures to later push the hydroturbine and convert osmotic power into electrical power. However, under high pressures, membranes inevitably lose selectivity because of deformation and damage of both TFC layer and supporting layer. In our work, a systematical investigation has been demostrated on the evolution of morphological deformation and membrane performance in terms of compacted structure, water flux, reverse salt flux, and power density as a function of hydraulic pressure for membranes used in high-pressure PRO processes. In addition, an insight to PRO membranes will be given on the effects of membrane modifications and post-treatments.

Acknowledgments

This research was funded under the project entitled ?Membrane development for osmotic power generation, Part 1. Materials development and membrane fabrication? (1102-IRIS-11-01) and NUS grant number of R-279-000-381-279. This research grant is supported by the Singapore National Research Foundation under its Environmental & Water Technologies Strategic Research Programme and administered by the Environment & Water Industry Programme Office (EWI) of the PUB. Special thanks are due to Dr. Panu Sukitpaneenit, Dr. Sui Zhang, and Miss Xiuzhu Fu for their valuable suggestions and kind help.

AUTHOR INDEX

AUTHOR INDEX

AUTHOR INDEX

AUTHOR INDEX